THE SOCIAL PSYCHOLOGY OF EXPERTISE

Case Studies in Research, Professional Domains, and Expert Roles

Expertise: Research and Applications

Robert R. Hoffman, Nancy J. Cooke, Anders Ericsson, Gary Klein, Eduardo Salas, Dean K. Simonton, Robert J. Sternberg, and Christopher D. Wickens, **Series Editors**

Hoc/Caccibue/Hollnagel (1995) • *Expertise and Technology: Issues in Cognition and Human-Computer Interaction*

Mieg (2001) • *The Social Psychology of Expertise: Case Studies in Research, Professional Domains, and Expert Roles*

Noice/Noice (1997) • *The Nature of Expertise in Professional Acting: A Cognitive View*

Salas/Klein (2001) • *Linking Expertise and Naturalistic Decision Making*

Schraagen/Chipman/Shalin (2000) • *Cognitive Task Analysis*

Zsambok/Klein (1997) • *Naturalistic Decision Making*

THE SOCIAL PSYCHOLOGY OF EXPERTISE

Case Studies in Research, Professional Domains, and Expert Roles

by

Harald A. Mieg
Swiss Federal Institute of Technology (ETH), Zurich

Psychology Press
Taylor & Francis Group
New York London

First published by
Lawrence Erlbaum Associates, Inc., Publishers
10 Industrial Avenue
Mahwah, New Jersey 07430

First issued in paperback 2012

This edition published 2012 by Psychology Press

Psychology Press
Taylor & Francis Group
711 Third Avenue
New York, NY 10017

Psychology Press
Taylor & Francis Group
27 Church Road, Hove
East Sussex BN3 2FA

Psychology Press is an imprint of Taylor & Francis, an informa group company

Cover design by Kathryn Houghtaling Lacey

Library of Congress Cataloging-in-Publication Data

Mieg, Harald A. (Harald Alard), 1961–
The social psychology of expertise: case studies in research, professional domains, and expert roles / Harald A. Mieg.
p. cm. – (Expertise, research and applications)
Includes bibliographical references and indexes.
ISBN 0-8058-3750-7
ISBN 978-0-415-65276-6 (Paperback)
1. Expertise. 2. Expertise–Social aspects. I. Title. II. Series.
BF378.E94 M54 2001
306.4'2–dc21

00-067766
CIP

Contents

Series Foreword

The first volume in this series, *Expertise and Technology* (J.-M. Hoc, P. C. Cacciabue, & E. Hollnagel, Editors), was an edited collection of reports on research involving cognitive modeling and cognitive systems engineering in various domains. The second volume, *The Nature of Expertise in Professional Acting* (T. Noice & H. Noice, Authors), described a program of psychological research on a particular domain. The third volume, *Naturalistic Decision Making* (C. E. Zsambok & G. Klein, Editors), presented reports on expertise from the perspective of the NDM paradigm. The fourth title, *Cognitive Task Analysis* (J. M. Schraagen, S. Chipman, & V. L. Shalin, Editors), presented a variety of studies of expertise, all illustrating the methods of cognitive task analysis. This, the fifth volume in the series, enriches the series by taking an integrative perspective. Along the lines of "cognition in the wild," Harald Mieg brings together considerations on expertise from psychology and sociology. To paraphrase William Mace's dictum for cognitive science, rather than looking just at what's inside experts' heads, Mieg looks at what experts' heads are inside of—organizations, social roles, management, and so on. Mieg illustrates his arguments using two case study domains: financial marketing and global climate change. We are very pleased to have Harald's work for the Series because it expands the horizons for expertise studies. It reinforces the notion that the full understanding of expertise in complex sociotechnological contexts requires researchers to bring to bear the methods and perspectives of sciences in addition to experimental psychology.

—Robert R. Hoffman

Preface

It has been said that experts are ubiquitous in today's life. We see them in courts, on TV, in hospitals, or as business consultants. Because experts are everywhere and are depended on so much, is it not ironic that social sciences have not, to date, thoroughly explored the question, who is really an expert? Most times it seems neither necessary nor possible to distinguish experts from other extraordinary persons such as professionals, specialists, scientists, or academics. Yet it seems an accepted fact that we cannot do without experts.

If questioned, most individuals would maintain that when being confronted with experts they may alternate between two attitudes: Sometimes they would like to consult the real expert—that is, somebody with superior knowledge who knows what to do and how the world functions. Sometimes they distrust this kind of objective advice and disregard expert knowledge as narrow-minded and being far from common sense. Some say: What we need is not certainty but useful knowledge.

Following this line of thought, this book looks at the use of the experts' expertise. Until today, the study of experts and their expertise has been covered by two sciences: psychology and sociology, precisely—the psychology of expertise and the sociology of professions. As I am trained in psychology and have a serious interest in sociology, I wanted to consider ways to bring together research from both disciplines. My first dialogues with scholars seemed promising because I heard the same remark on both sides: "Interesting!" However, what I eventually realized was that "Interesting!" never marked the beginning of discussion, but the polite end. I was confronted with a gap in a scientific *weltanschauung* (worldview). From the psychological point of view, experts are defined by intrinsic characteristics; experts are persons with special knowledge or capabilities. From the sociological point of view, experts stand for specific qualifications and social status, with individual differences being irrelevant.

I believe that both perspectives on experts are inherently true, albeit too simple an answer. It can hardly be denied that experts are human beings; thus, the specific limitations of human cognition shape the use of all kinds of experts—an argument toward sociology. Neither can this be denied: That which counts as valuable expertise depends on social definitions. Thus, expertise cannot be explained by intrinsic factors only—an argument toward psychology. To display these arguments, I have previously written a scientific thesis, "Expert Thinking and the Expert's Role" (a so-called *habilitationsschrift*), and now I introduce this volume, *The Social Psychology of Expertise.*

I should add that this volume bears the imprints from my work at the Swiss Federal Institute in Zurich, which results from daily interaction with engineers and scholars (experts) from natural sciences and economics. Much of this work consists of explaining social science to technical experts. My book is a journey through research in psychology and sociology, and it provides case studies in financial markets as well as climate change. It concludes with an outlook on the managerial problems with experts. I would like to foster an understanding of what experts do and how we can make proper use of their expertise. I would like this book to contribute to a mutual understanding between the psychology of expertise and the sociology of professions, providing concepts that support the scientific knowledge exchange for research on expertise.

—Harald A. Mieg

Acknowledgments

This book is indebted to the support and helpful comments of many scholars and friends. Unfortunately, it is impossible to name all of them. I definitely have to thank Andrew Abbott, Chicago; Michael Corsten, Berlin; Olive C. Crothers, New York; Erwin Hepperle, Zurich; Carlo C. Jaeger, Potsdam; Helmut Jungermann, Berlin; Pieter Leroy, Nijmegen; Volker Lipp, Mannheim; Andreas Oehler, Bamberg; Atsumu Ohmura, Zurich; A. J. M. Schoot Uiterkamp, Groningen; and Rudolf Volkart, Zurich. Particularly, I have to mention Roland W. Scholz, Zurich, who taught me to do science from an interdisciplinary point of view; and Robert R. Hoffman, Pensacola: Thanks to him, this volume has come into being.

CHAPTER

1

Introduction

> *One of the gravest of cognitive problems in the modern world is that of rendering accessible in an organized, coherent, and coordinated way the information already, broadly speaking, available—a process that is invariably difficult and expensive.*
>
> —Nicholas Rescher (1989, pp. 10–11)

In a nutshell, this quote from philosopher Nicholas Rescher shows many of the issues regarding the dissemination and use of knowledge in today's knowledge-based societies:

- First, the use of knowledge requires an institutionalized form of exchange (organized way), such as books, schools, or experts.
- Second, knowledge is time-dependent; relevant information refers to the knowledge base of a particular time (information already available).
- Third, exchanging knowledge can be costly (difficult and expensive); as far as knowledge is subject to valuation and selection, it has an economic dimension.

This book focuses on experts as part of a society's knowledge base. It presents cognitive expertise as a particular sort of an individual, human capacity. Expertise is based on knowledge. But is it the same? There are doubts. Expertise has its specific developmental aspect—we have to train to become experts. In contrast, knowledge per se seems to have an impersonal, transpersonal quality. Speaking of cognitive problems, Rescher, as a philosopher, did not refer to individual problems of cognition, but to a general problem of rationality. The general question "How can cognition be considered both human *and* rational?" is remarkably fundamental. In this book, we can only touch it but not thoroughly explore it. Instead, we look at human expertise as the basis of a society's knowledge and its exchange.

The title of this book—*The Social Psychology of Expertise*—reflects an approach where we will not speak of expertise except in the social context. Thus, the book is concerned with experts in their working contexts and social interactions—this includes everyday phenomena such as

- disputing experts,
- experts who err,
- rather limited expertise or limits to experts, respectively, and
- experts who cannot clearly explain what they do.

However usual these phenomena are, they seem to conflict with our expectations about what experts should be. In this book, we enter a discussion on the role of experts and on rendering cognitive expertise accessible—a discussion similar to the discussion on knowledge exchange with Rescher. From the point of view of social psychology, we consider both the inner, psychological side of expertise and its outside—the social function of expertise. In particular, we have to ask: "How can knowledge exchange be considered both an instance of individual expertise and the realization of a social function?"

Obviously, expertise concerns a person and a function. The remainder of this chapter acquaints the reader with:

- how our question is linked to the discussion of expertise in current psychology,
- how we can tackle it from a social psychology point of view, and
- the rationale and content of *The Social Psychology of Expertise*.

1.1 EXPERTISE

What is an expert? Experts—in the original literal sense—are experimentalists: They know from active, reflexive experience. Accepting this definition as a starting point, we can ask: What is special as to experts? Does not everybody know from—more or less reflexive—experience? *The Nature of Expertise*, a textbook on the cognitive psychology of expertise edited by Chi, Glaser, and Farr (1988), started with the introductory question, "How do we identify a person as exceptional or gifted?" (Posner, 1988, p. xxix). This kind of question leads us into a *differential* approach, comparing experts with nonexperts. From a psychological point of view, there are two further directions. First,we can look for differences in personality: where experts excel in intelligence, reasoning strategies, or cognitive information-processing capabilities. Second, we can look for differences in learning conditions such as training and schooling or cognitive stimulation. In addition, we can mix both approaches and describe expertise as the result of a specific developmental, learning-based process that shapes a personality—the expert.

We find this differential approach with K. Anders Ericsson. In his book, *Towards a General Theory of Expertise* (edited with J. Smith in 1991), he wrote,

> [T]he study of expertise seeks to understand and account for what distinguishes outstanding individuals in a domain from less outstanding individuals in that domain, as well as from people in general. (Ericsson & Smith 1991a, p. 2).

Ericsson limited his approach to the study of cases, "in which the outstanding behavior can be attributed to relatively stable characteristics of the relevant individuals" (loc. cit.). Ericsson argued, "We believe that stability of the individual characteristics is a necessary condition for any empirical approach seeking to account for the behavior with reference to characteristics of the individual" (Ericsson & Smith 1991a, p. 2).

Paradigmatic examples for this kind of approach are excellence in chess, sports, and music. In these cases, we have competitions with high rewards and established performance criteria. Moreover, there seems to be a clear line of development with a long phase of individual training and growing expertise, distinctive phases of superior performance, and, generally, some kind of retreat or final resignation. Following Ericsson, only some of the individuals in these fields are experts—the best ones. The question that remains is: How representative are these sorts of expertise for expertise in general? Do we think of competitions of chess players, athletes, or musicians when speaking of experts' disputes? Ericsson's approach risks excluding from analysis many interesting cases of expertise from the beginning.

The textbook *The Nature of Expertise* presented examples of expertise from a variety of fields, including typewriting, restaurant orders, mental calculation, computer programming, judicial decision making, and X-ray diagnostics. In many cases, the experts were the individuals with high task performance. In other cases, they were professionals such as physicians. If we apply Ericsson's expert criterion to professionals, we have to ask: Is every professional an expert showing outstanding performance? Are there professions without any experts or expertise? Lacking overt competition criteria, the studies in fields such as medical diagnosis do not refer to differences between single professionals, but between professionals and non-professionals or professionals and students. Shifting from an expert definition based on outstanding performance (such as sports champions) to a definition based on professionalism, the understanding of expertise becomes dependent on an analysis of professional work.

Sylvia Scribner went one step further in her studies in 1984. She examined expertise in blue-collar work in a medium-sized milk-processing plant. Her subjects were, for instance, preloaders and wholesale drivers, but also some clerks. She invented experimental cognitive tasks parallel in structure to specific everyday work in the plant. She could demonstrate that any group of workers outperformed the other groups in tasks that had a cognitive structure similar to its everyday work. For instance, preloaders responsible for the assembly of milk cases showed "a large repertoire of solution strategies" for product assembly tasks (p. 21). Scribner spoke of *working intelligence* and concluded that: "expertise is a function of experience" (p. 24)—a conclusion that is a general working hypothesis in cognitive psychology. Expertise is mainly based on *experience*. This has also been the intent of the often cited verse "Experto credite" by the Roman poet Virgil: Trust the one with personal experience. Our question (What is special as to experts?) can now be provisionally answered: It is superior performance based on specialized experience.

A completely different concept of expertise, although similar at first glance, is to understand experts as specialists having specialized knowledge. Whereas the concept of expertise-by-experience views expertise from an inner, cognitive point of view, the reference to knowledge starts with an outside view of expertise. Here, again, we see a difference between knowledge and experience. An expert-by-experience must be an expert in a field. An expert-by-knowledge can also be an expert about the field, lacking personal experience in the field. Notoriously, this is the case in academia. In general, academics are experienced in academic life, including adaptation to academic performance criteria. However, academic and scientific knowledge generally refers to phenomena outside the university—nature, society, human health, and so on. Historians, archaeologists, astronomers, and other scientists do not even have a real chance to physically enter the field of their scientific concern.

The understanding of experts as knowledge specialists meets everyday concepts—prejudices—about experts. In 1990, Donald N. McCloskey, a scholar in history and economy, described the "narrative of economic expertise." His book's title expresses what McCloskey called the "American question: If you're so smart why ain't you rich?" If someone as an expert has sound knowledge about economics, he or she should also be able to utilize this knowledge in real business. This is not generally the case. McCloskey cited proverblike remarks on experts that reflect this concept of experts as knowledge specialists and expresses one part of public opinion on expertise:

> Harry Truman: "An expert is someone who doesn't want to learn anything new, because then he wouldn't be an expert." (see McCloskey, 1990, p. 111)
>
> N. M. Butler, former president of Columbia University: "Experts know more and more about less and less." (loc. cit.)

McCloskey (1990) resumed in the same direction: "The expert as expert, a bookish sort consulting what is already known, cannot by his nature learn anything new, because then he wouldn't be an expert. [. . .] Smartness of the expert's sort cannot proceed to riches" (p. 134). Can we leave aside the concept of expert-by-knowledge? It might be a viable strategy of psychology to discuss no cases of experts with uncertain expertise, but focus on analyzing real expertise that fulfills high-performance criteria. This strategy might also fulfill the wish of the public to sort out bad expertise. However, there is a serious phenomenon to explain so-called experts that lies beyond the explanatory range of defining experts by performance. These are, for example, experts giving political advice or consulting international firms. We cannot understand these cases if we do not take very seriously what Hoffman, Feltovich, and Ford (1997) concluded, resuming the state of the art of the psychology of expertise—namely: the "minimum unit of analysis" is the "expert-in-context" (p. 553).

1.2 SOCIAL PSYCHOLOGY

Social psychology is concerned with the interactions between persons and situations, the situations being defined by, for instance, groups, organizations, or personal relationships. This also pertains to experts. Irving L. Janis (1972), in his book *Victims of Groupthink*, analyzed failures of expert advise caused by the dynamics of an expert advisory group. The centerpiece of his book is the analysis of the Bay of Pigs invasion. On April 17, 1961, a trained group of about 1,400 Cuban exiles, aided by the CIA, U.S. Navy, and U.S. Air Force, invaded the coast of Cuba at the Bay of Pigs. However, nothing went according to plan. Within 3 days, all invaders were killed or captured by Cuban troops. The mission ended as a complete fiasco.

President John F. Kennedy, who decided on the invasion, was advised by a group of highly qualified experts. Nevertheless, the group—including Kennedy himself—made assumptions that proved to be completely wrong. For example, they mistakenly assumed that the invasion would provoke armed uprisings in Cuba. Janis described this phenomenon as a groupthink syndrome, consisting of

- overestimations of the group—its power and morality,
- closed-mindedness, and
- pressures toward uniformity.

The expert advisory group did not even realize that one central element of the plan disappeared due to an alteration in other parts of the plan: The invading exiles should at least have had the opportunity to retreat in the Cuban mountains—an impossibility after having finally decided to land at the Bay of Pigs.

Janis' analysis pertains to policymaking groups in general and seems not specific to expert advise. Unfortunately, the analysis gives no answer to questions concerning the expertise of the involved advisors: How did their role as experts or specific aspects of their expertise contribute to the result of the expert advisory group? How is expertise linked to the interaction as expert?

We find more studies in this direction in the sociology of science. For instance, H. M. Collins (1985) studied replication and induction in scientific practice. In particular, Collins found an *experimenters' regress*: "[S]ince experimentation is a matter of skillful practice, it can never be clear whether a second experiment has been done sufficiently well to count as check on the result of the first. Some further test is needed to test the quality of the experiment—and so forth" (p. 2). In a thorough case study, Collins described the replication of the Transversely Excited Atmospheric (TEA) Laser—a special laser that uses carbon dioxide at atmospheric pressure. Collins showed that even when every component is explicitly known, building the laser is a skill derived from training and developed over time. Collins (1985) concluded, "Experimental ability has the character of a skill that can be acquired and developed with practice. Like a skill, it cannot be fully explained or absolutely established" (p. 73). The experimenters' regress is basically connected to peculiarities of scientific expertise. It takes time to become capable of properly conducting a scientific experiment. From this point of view, it is not surprising to see science divided in schools that develop their specific methods and adhere to diverging theoretical frameworks.

A similar concept of developing knowledge, but that is far more generalized, is proposed by William J. Clancey. In his approach to *situated cognition*, he tried to "re-relate" human knowledge and programs of artificial intelligence (AI; Clancey, 1997b). Clancey claimed that human activity, including knowledge, basically routes in an adaptation to environmental constraints:

> Every human thought and action is adapted to the environment, that is, *situated*, because what people *perceive*, how they *conceive of their activity*, and what

> they *physically do* develop together. From this perspective, thinking is a physical skill like riding a bike. (Clancey, 1997a, pp. 1–2; italics original)

This also means that knowledge—even scientific knowledge—has to be seen in the context of social activities. "To understand the idea that *knowledge* is inherently social (as well as inherently neural), we must first understand that human *action* is inherently social" (Clancey, 1997b, p. 263). From the perspective of situated cognition, knowledge is instrumental; it is a means for social activity. "An individual's capacity to engage in an activity may be characterized as knowledge. Thus, 'knowledge is socially constructed' means first, that knowledge develops and has value within activity, and second, activities are socially constructed" (Clancey, 1997b, p. 270).

With his approach, Clancey led us into epistemological questions of the nature of knowledge. Clancey referred to the American philosopher John Dewey who considered representations as tools for inquiry (Clancey, 1997b, p. 259), regardless of whether the representations were external (e.g., charts, books) or internal (e.g., silent speaking, thinking). The same idea was expressed by the philosopher Ludwig Wittgenstein in his remarks on concepts as *instruments*: "Language is an instrument. Its concepts are instruments. . . . Concepts lead us to make investigations; are the expression of our interest, and direct our interest" (*Philosophical Investigations*, n.d., no. 569 & 570).

Speaking of experience in the framework of situated cognition, we have to be aware that experience now means something different than experience from active individual learning. According to the situated cognition approach, experience strongly depends on a physical environment; in particular, it involves the use of instruments and artifacts such as books, computer programs, tape measurers, ammeters, or scientific language. Expertise cannot be as easily attributed to stable characteristics of an individual. Instead, we have an expert-in-context, such as a researcher with his lab where he tries to replicate a TEA Laser. Then an increase in experience means an increasing degree of adaptation to a specific working setting. Hence, Butler was quite right in saying that experts know more and more about less and less.

In his case study *Cognition in the Wild*, Edwin Hutchins (1995) showed the role of mediating situational structures of which situated cognition consists. Hutchins, scientific guest on a U.S. Navy ship, was present when the ship's propulsion system failed during an entry into San Diego Harbor. Because there was shoal water on one side of the sweeping ship and a shipping channel on the other, it was important to maintain awareness of the location of the ship. A quartermaster of the navigation team had to directly measure with a pelorus—a navigational instrument—the direction of the landmark with respect to the ship's head. In cooperation with a second team member, the true bearing of the landmark had to be continuously computed. Hutchins analyzed the series of computations of changing bearings. He showed how the two team members arrived at several stable solutions for the calculation problem, particularly "how the behavior of one individual can act as mediating device that controls the patterns of availability of data for the other" (p. 340).

In this framework, Hutchins also provided three explanations of why we often cannot say what we do—or why experts sometimes cannot say what they do:

- first, remembering is a kind of coordination that is different to the one of original action,
- second, reports are often based on control structures (that differ from the skills to be controlled), and
- third, the mediating structure is only present in the environment. (pp. 310–311)

These examples show the explanatory potential of the situated cognition approach for the explanation of expertise. However, this approach still focuses on the cognitive interpretation of a problem that one individual—the expert—has to solve aided by others or by mediating situational structure. To complete this picture of expertise-in-context, we should take into account the cognition of the person who would like to have solved the problem by this particular expert. Equally, we have to take into account the meditating structures that help identify the particular expert.

Agnew, Ford, and Hayes (1997) put the provocative question of why we deny expert status to snake oil salesmen, TV evangelists, and chicken sexers when we grant it to geologists, radiologists, and computer scientists:

> What do snake oil salesmen, TV evangelists, chicken sexers, small motor mechanics, geologists, radiologists, and computer scientists all have in common? They all meet the minimum criterion of expertise, namely they all have a constituency that perceives them to be experts. (Agnew, Ford, & Hayes, 1997, p. 219)

Moreover, they insisted that *expert* denotes a *role* "that some are selected to play on the basis of all sorts of criteria, epistemic and otherwise" (p. 220). There are, they added, "many niche-specific characteristics and performance criteria" (loc. cit.).

We see that we still have to take the step from describing singular expertise to explaining experts-in-contexts. Similarly to speaking of knowledge as an instrument, we could conceive of experts as instruments we can use. It is one aim of this book to examine the specific uses of experts and the attributions that lead us in identifying them.

1.3 THE RATIONALE OF *THE SOCIAL PSYCHOLOGY OF EXPERTISE*

The rationale of this book is threefold: It provides practical insight, there is a necessity to connect research, and we should understand the methodological need for scarcity. The practical insight is simply that there are relative experts; that is, persons who are used as experts only in a particular context. The insight also shows that relative expertise is a common feature. That is to say, it is so common in our everyday lives that we do not realize it. The necessity to connect research concerns research on experts and expertise. In the social sciences, there are two traditions of studying experts: the psychology of expertise and the sociology of the professions. These parallels would seem to be a perfect case for interdisciplinary research that could also contribute to an understanding of the human component in the knowledge base of modern societies. As a matter of fact, however, this knowledge exchange does not take place. One obstacle toward an integrated view is the ignorance of fundamental differences among the disciplinary views. Therefore, *The Social Psychology of Expertise* pleads for methodological scarcity. These three aspects—practical concern, research, and methodology—define three paths through this book.

The Practical Insight: Relative Expertise

Relative expertise is a useful, common feature. In every family, there is a shared knowledge about who knows what and whom you have to ask; for instance, where to find the clean bedding, what to do in case of a blocked drain, or whom to call if the TV set breaks down. The same is true among friends and colleagues. Similarly, in every organization, there is always one person who—without any authority—seems "to know everything and everyone" and whom we can ask anything regarding the organization's informal structure. Relative expertise is not only related to practical issues of everyday knowledge, but also to scientific research and essential questions of life. The borders between relative expertise and science or professional work are fuzzy, in particular when among your friends you have scientists or professionals whom you can consult. Other instances would be if a student is focused in a particular subject, a journalist is writing about it, or you consult a friend who has read all the books by this journalist.

Thus, we have to define the context of using experts and expertise: This includes the ones who make use of others' expertise (clients and audience), the kind of problems to be tackled, and the role of knowledge (scientific). Chapter 4 examines expertise in the organizational context—in particular, the role conflicts with experts. We can take on a sociological view and try to explain these role conflicts on the level of professionalism: Conflicts are attributed to an incompatibility between autonomy of single professionals and bureaucratic constraints of organizations. From this perspective, professionals are trained to work as their own masters and face problems when being fitted into a firm's hierarchy. As a matter of fact, on that level, the role conflicts empirically disappear. However, there are other sources of role conflicts, especially in teams. "Who has to explain what?" is a question that drives group dynamics and also defines responsibilities. This can lead to misjudging expertise in teams as well as in organizations. We see how role conflicts with experts can be explained and resolved by understanding them as resulting from "The expert"-interaction (which is introduced in chap. 3).

Conclusions for the management with experts are contained in chapter 8. The problem of management starts with the fact that it is related to the control of human behavior. Thus, management is linked to uncontrollable feedback effects that render it a weak form of expertise. Chapter 8 provides an expert role approach for managerial planning that connects a typology of experts (from chap. 7) with an analysis of the sequence of professional work. The expert role approach provides an idea about effective decision support through types of experts. As one vision, chapter 8 proposes scenario-based planning to profit from a differentiated integration of experts and expertise with managerial decision making.

The Necessity to Connect Research: Psychology of Expertise—The Sociology of the Professions

As said, the knowledge exchange between the psychology of expertise and the sociology of the professions currently does not take place. It is not a lack of mutual interest, it is—inter alia—a difference in points of view or, to put it another way, an epistemological problem. The working hypothesis for psychologists is like this: Knowledge is an objective quantity that can be implemented in individuals. Sociologists consider knowledge as relative to a social construction of reality. This book, *The Social Psychology of Expertise,* picks up both lines of research. It draws much from the outstanding disciplinary work on both sides—in particular, Andrew Abbott's (1988) *System of Professions*, which fits perfectly into research on expertise in cognitive psychology. *The Social Psychology of Expertise* provides an interdisciplinary, integrated view on experts and invites the reader to ponder the use of experts in professional settings.

Chapter 3 contains the essentials for understanding expertise-in-context. This chapter introduces a concept of the expert's role that shall provide a conceptual bridge between the psychology of expertise and the sociology of the professions. As proposed in chapter 2: "The expert" is not a type of person, but a form of interaction involving an attribution to a person—a social form. The attribution is relational, such as *neighbor* or *respondent*, and helps identify a person as having a specific cognition capacity. Usually in "The expert"-interaction, someone explains something to someone else (but this is not a necessary feature). Our definition does not aim at a full description of the phenomenon called *experts*. On the contrary, it provides a minimal explication that makes it possible to empirically investigate many of the alleged characteristics of experts and their roles: How reliable is the expert's opinion? What is its basis in experience and science? Do experts serve to legitimize political interests?

Chapter 3 also discusses the role of knowledge. We see that knowledge is neither data nor information about facts. Knowledge essentially needs interpretation. Accordingly, we can understand experts as knowledge interpreters. They interpret a problem or question from the point of view of knowledge that is valid within a particular knowledge community such as science or a profession.

Finally, chapter 3 is concerned with the performance aspect of expertise. Now experience comes into play—along with the notion of *truth* (a concept much in dispute in social sciences). When consulting an expert, we might doubt the value of the explanations and knowledge the expert provides. However, in general, we do not suspect that this is the knowledge the expert can provide; therefore, the expert is often identified with his or her knowledge. Truth is a presupposition of "The expert"-interaction. It does not refer to absolute knowledge, but to objectivity in the sense of social generalizability: If we had the time to have the same experiences as the expert, we would come to the same conclusions. Chapter 3 provides the central proposition of *The Social Psychology of Expertise*: It is the time gain for what we are willing to pay in "The expert"-interaction or it is the relatively fast utilization of the expert's compressed experience that any reasonable person could make if she or he had enough time to do so.

Chapters 5 and 6 present two case studies on the relationship between expert roles and expertise-in-context. Such studies require one to view the whole context. The cases are two human–environment systems—financial markets and climate change research, both of which are domains of strong public interest but uncertain expertise.

Chapter 5 is concerned with experts in financial markets. From the point of view of expertise, financial markets can astonish because:

- there is established theory (about finance, markets, risk analysis), and
- there are continuous measurement and clear performance criteria (e.g., rates of returns).

Nevertheless, expert performance in forecasting financial markets is low. The case of forecasting financial markets shows an example of weak-form expertise due to systemic market effects, some of them being unexpectedly caused by expert advice.

Chapter 6 displays the case of research on climate change between 1988 and 1997. Predicting climate change has to take into account so many uncertainties, including methodological ones, that finding uncertain expertise and dispute among experts could not really be surprising. The effects of climate change are global and can only be assessed on a long time scale—reaching after the death of most of the experts. Surprisingly, in dealing with climate change as well as in environmental issues in general, a sort of nonscientific lay expert comes into play to support science. We call them *system experts*; they are the individuals who know well the local conditions of the human–environment system in which they live.

Chapter 7 presents the conclusions from the case studies for the conceptualization of expertise-in-context. The starting point is the fact that not every kind of uncertainty creates a demand for experts. No one worried about computers before World War II—or in the Middle Ages. In particular, the societal or personal need for control results in a demand for experts. Therefore, we have to separate the epistemological question (What do we know?) from the pragmatic question (What shall we do?). Chapter 7 introduces a distinction between uncertainty—as lack of knowledge—and insecurity—a need for control of uncertainty. Furthermore, Chapter 7 designs a typology of experts depending on the knowledge they administer. This typology—as well as the other concepts in this concluding chapter—should be seen in the way Dewey and Wittgenstein conceived of concepts in general: as an instrument for further investigation. Consequently, chapter 7 ends with a view of experts as heuristics—that is, instruments for short-cutting problem solving.

Methodology: Occam's Razor

Occam William from Ockham in the county of Surrey was a philosopher of the late Middle Ages. Occam is famous for having postulated a principle of ontology (the science of the being): *Entia non sunt multiplicanda sine necessitate* (Entities must not be multiplied without necessity). This principle is called *Occam's razor* because it should restrain us from overconceptualization. Therefore, it is not surprising that the theoretical framework of *The Social Psychology of Expertise* is not characterized by its richness, but its scarcity. To put it metaphorically: To build a bridge, we do not need to build a city—that would in toto incorporate the boundary to be crossed by a bridge.

The framework—introduced in chapter 1—is called *cognitive economics*, but it is much less than a theory in economics or cognitive psychology. It owes much to Herbert A. Simon, who coined the term *bounded rationality* and defined a way of thinking that deeply influenced both economics and psychology. Cognitive economics refrains from presuppositions about rationality. Instead, its basic proposition is the purely quantitative limitation of individual human cognition. No one knows everything. There are cognitive and societal mechanisms to overcome this limitation, heuristics being a cognitive one, experts a societal one. We assume this: If there is a societal demand for cognitive capacity, there is also a market for it. The social allocation of cognition capacity (e.g., via expert roles) affects the way in which societies' problems are solved. Thus, the specific roles of experts and expertise, respectively, reflect an important part of a society's productive knowledge base.

CHAPTER

2

Where We Should Start: Cognitive Economics

This chapter introduces a framework entitled *cognitive economics*. The core premise is the purely quantitative limitation of individual human cognition. There are cognitive and societal mechanisms to overcome this limitation. We see how experience-based expertise economizes information-storing capacity as well as processing time. Cognitive economics assumes that if there is a societal demand for cognitive expertise, there is also a market for it. For instance, professions are one form of regulating economic transactions related to human expertise. Thus, experts have to be seen in connection with a demand for expertise.

2.1 ECONOMIZING INFORMATION-STORING CAPACITY: THE MAGICAL NUMBER SEVEN, PLUS OR MINUS TWO

In 1956, the *Psychological Review* published an article with the wonderful title: "The Magical Number Seven, Plus or Minus Two." George A. Miller, the author, wrote that he had been obsessed by this integer, the number 7, for years. "It has intruded in my most private data, and has assaulted me from the pages of our most public journals," (p. 81). His discovery, if we speak in today's terms, was that the short-term memory has a loading capacity of around seven units, plus or minus two. The consequences of this single discovery were enormous.

From Bits to Chunks

To understand Miller's finding, we have to consider the state of the art of psychology at that time. After World War II, a new paradigm arose: information theory. During World War II, information theory had helped decode radio messages by disentangling messages and noise. In the information theory model, information that is to be sent from the source—perhaps a person—to the destination—perhaps another person—has to be encoded, passed through a channel, and finally decoded. For example, if one person, say John, wants to greet another person, we call her Brenda, he has to encode his message. He might say, "Hi Brenda!" The message passes through the air and is mixed with considerable noise, say from a passing van. Finally, the message has to be decoded by Brenda.

The psychological research to which Miller referred regards the person as a channel. His research was designed to present pure informational inputs—sounds, tastes, colors—and measure how much of the information the subject could recognize and recall. Information is measured in bits. One bit is the information required for a simple binary decision: For example, a switch can be on or off. To know whether this switch actually is on or off, it must contain one bit of information. If we have two switches (two bits of information), there are four possible combinations. The general formula is $N = 2^k$, where N is the number of alternatives and k is the information measured in bits. Measuring information with bits does not include anything about the contents of the information. If we have a system of eight colors, the information is 3 bits ($2^3 = 8$). Three bits is also the information necessary to denote one of eight neighbors or, as in our first example, the positions of three switches.

Miller discovered that in one-dimensional stimuli—such as sounds, tastes, and colors—subjects could identify between 1.6 and 3.9 bits of infor-

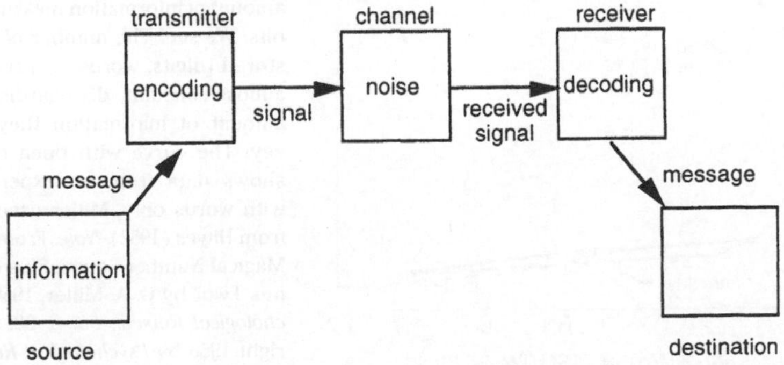

FIG. 2.1. The channel model of information theory (see Shannon & Weaver, 1949).

mation. The studies Miller reanalyzed showed an average of 2.6, corresponding to about 6.1 categories. In the terms of information theory, the human channel capacity for one-dimensional stimuli is about 2.6 bits; but this is only half the story. Consider, for instance, a telephone number that consists of eight digits:

1 9 5 3 1 9 5 6

The total amount of information conveyed by these eight digits is 26.6 bits. Perhaps you can manage to keep the eight single digits in mind. However, if you imagine the digits as two dates,

1953 1956,

you will more easily remember them. Thus, not the number of bits but the number of units needed to recall constrains your recall capability. Miller spoke of *chunks* of information. Disregarding the quality and complexity of the chunk, the memory span is equal to seven units. These may be seven single digits, seven dates, or even seven familiar telephone numbers.

Today, Miller's article is regarded as classic general psychology. We now speak of short-term memory rather than human channel capacity. The short-term memory holds information for no longer than about 20 seconds and forms one part of our memory system that also encompasses the sensorial or ultrashort term memory (2 seconds) and the long-term mem-

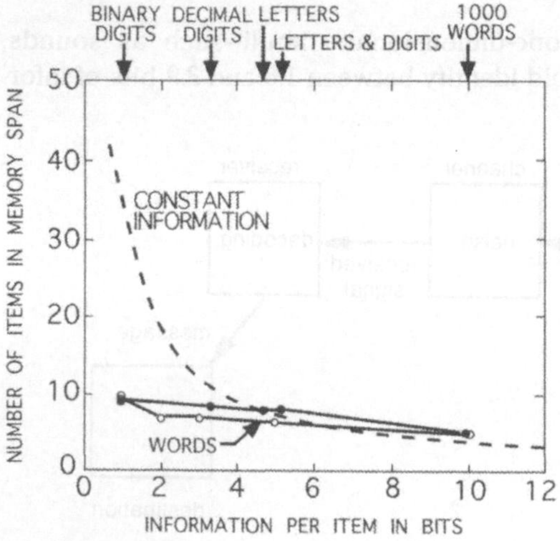

FIG. 2.2. The magical number seven, plus or minus two. The solid curve shows experimental data, indicating numbers of items [chunks] in the memory span [short-term memory] as a function of the amount of information measured in bits. We see: The number of items stored [digits, words . . .] remains almost constant, disregarding the amount of information they convey. The curve with open circles shows data from an experiment with words only. Miller used data from Hayes (1952). *Note.* From "The Magical Number Seven, Plus or Minus Two" by G. A. Miller, 1956, *Psychological Review, 63*, p. 92. Copyright 1956 by *Psychological Review.* Adapted by permission.

ory. Rehearsal, meaning the act of mental recalling, is needed to keep chunks within our short-term memory.

The Mind's Eye in Chess

The concept of chunking offers an understanding of the remarkable feats of expertise as frequently seen in chess. Pioneering work was conducted by Adriaan D. deGroot, a master player who systematically investigated the cognitive processes underlying chess mastery as early as the 1940s. His basic question was this: What is it that differentiates master players from weaker players, even expert players? He compared the cognitive skills and processes of chess masters to those of beginners and advanced players. deGroot studied some of the best chess players in the world.

The way deGroot proceeded was to show chess players a chess position and ask them to find the best move. The study had two groups: beginners and masters. During the study, the players had to talk aloud while thinking. deGroot (1965) wrote: "In view of the large difference in playing strength we must assume that differences exist between the two groups" (p. 319). One common hypothesis was that masters consider more possible moves than beginners do. However, the analysis of the verbal protocols revealed no gross differences in thought processes. Masters search about as deep as weaker players do. "It is not generally possible to distinguish the protocol of a grandmaster from the protocol of an expert player solely on structural and/or formal grounds" (loc. cit.). Astonishingly, the chess masters consider fewer alternatives than weaker players do before choosing a move.

According to another common hypothesis, masters, not using special cognitive strategies, may dispose of special cognitive competencies. For instance, do chess masters have enlarged memory capacities? deGroot conducted a series of experiments involving perceptual and short-term memory processes. For an exposure time of 2 to 5 seconds, a chess position

FIG. 2.3. Chess position typical to deGroot's experiments. *Note.* From *Thought and Choice in Chess* (p. 326) by A. D. deGroot, 1965, The Hague: Mouton. Copyright 1965 by Mouton. Reprinted by permission.

TABLE 2.1
Results of Series of 10 Selected Positions for
Two Master Players and Two Advanced Chess Players

	Mean Exposure Time (seconds)	*Mean % Correct*	*Mean % Defect*	*Perfect Reproductions*
Master	3.2	93%	7%	4 of 10
Master	3.2	93%	7%	2 of 10
Advanced player	3.65	72%	28%	none
Advanced player	3.85	51%	49%	none

Note. From *Thought and Choice in Chess* (p. 329) by A. D. deGroot, 1965, The Hague: Mouton. Copyright 1965 by Mouton. Reprinted by permission.

chosen from "relatively obscure actual master games" was presented. Subjects were requested to dictate the position from memory. Masters were able to reproduce almost all positions correctly. However, there was a gulf between the masters and the nonmasters. The masters in deGroot's experiments in the long run showed a mean defect rate of less than 10%, whereas the defect percentage of the nonmasters ranged higher than 30%—even as high as 50%. deGroot concluded that even if masters do not calculate more than weaker players, they definitely see more.

What do the results prove concerning the memory capacities of chess masters? To determine the answer, deGroot chose a variation of his first experimental design. The chess pieces were randomly placed on the chess board. This time, the recall capability of masters and nonmasters did not differ. Recall was equally poor for masters and weaker players. Thus, masters are subject to the same memory limitations as everyone else.

In a series of experiments, William G. Chase and Herbert A. Simon (1973) at Carnegie Mellon confirmed deGroot's results. They spoke of the *mind's eye in chess*, concluding that "the most important processes underlying chess mastery are these immediate visual-perceptual processes rather than the subsequent logical-deductive thinking processes" (p. 215). They tried to explain the "feat of memory" of chess masters and studied what it is that chess masters perceive when exposed to a chess position. They discovered that masters perceive *meaningful constellations*: "It appears that the master is perceiving familiar or meaningful constellations of pieces that are already structured for him in memory, so that all he has to do is store the label or internal name of each such structure in short-term memory" (p. 217).

Thus, chess constellations define cognitive, informational chunks in exactly the sense described by Miller. Modern research on chess expertise specified the type of chunks relevant to chess mastery. We cannot consider chunks in chess as perceptual patterns of chess pieces. Rather, there is a

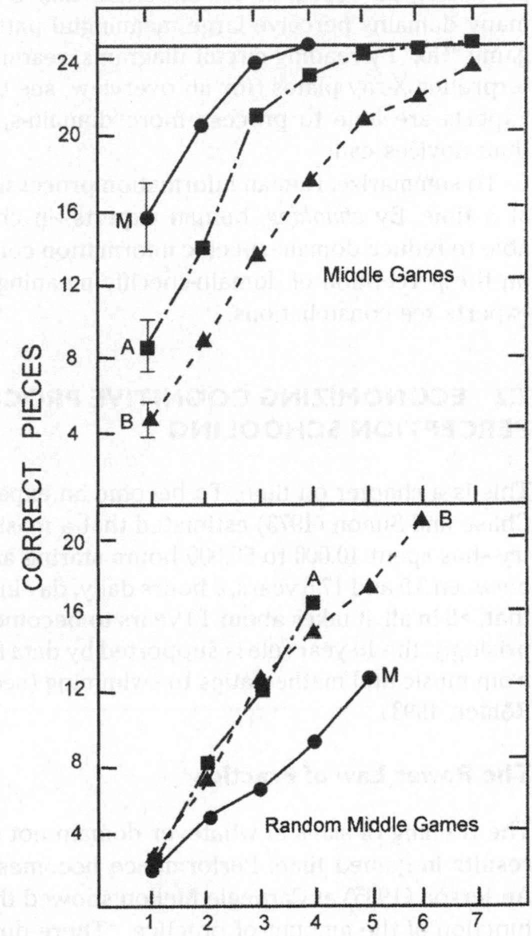

FIG. 2.4. Chess masters, beginners, and random positions. Number of pieces correctly recalled over trials for real "middle games" and randomized "meaningless" positions. M: master, A: advanced player, B: beginner. [The brackets on Trial 1 represent one standard error, based on five positions.] *Note.* From *Visual Information Processing* (p. 221) by W. G. Chase, 1973, New York: Academic Press. Copyright 1973 by Academic Press. Reprinted by permission.

"balance between knowledge and search" (Charness, 1991): Chess masters perceive chess constellations as meaningful *moves* or strategic positions in a game (Freyhof, Gruber, & Ziegler, 1992). Therefore, the masters' feat of memory disappears once the chess pieces are randomly placed on the board, and no meaningful constellations are available.

deGroot's findings are central to the psychological investigation of human expertise. After deGroot—and William G. Chase—the comparison of experts and novices became a standard research design. Experts are the masters of a domain of specialized skills or knowledge; novices are advanced pupils whose aim is to become experts. We have to distinguish them from lay persons, who have no experience with the domain. In the last 20 years, psychological research on expertise has demonstrated that experts in many domains perceive large meaningful patterns, such as in the Chinese game “Go,” by reading circuit diagrams, reading architectural plans, and interpreting X-ray plates (for an overview, see Chi et al., 1988). In a nutshell: Experts are able to process more domain-specific information at a time than novices can.

To summarize: Human information processing is limited to seven chunks at a time. By *chunking*, human experts—in chess and other domains—are able to reduce domain-specific information complexity. This ability is based on the perception of domain-specific meaningful patterns. In other words, experts *see* constellations.

2.2 ECONOMIZING COGNITIVE PROCESSING TIME: PERCEPTION SCHOOLING

This is a chapter on time. To become an expert on chess, one needs time. Chase and Simon (1973) estimated that a master—1 of the top 25 of a country—has spent 10,000 to 50,000 hours staring at chess positions; this means between 3.5 and 17.5 years, 8 hours daily, day in and day out. They concluded that, all in all, it takes about 10 years to become a real chess expert. Not surprisingly, the 10-year rule is supported by data from a wide range of domains: from music and mathematics to swimming (see Ericsson, Krampe, & Tesch-Römer, 1993).

The Power Law of Practice

The training of skills in whatever domain not only consumes time but also results in gained time: Performance becomes increasingly faster. John R. Anderson (1985) at Carnegie Mellon showed that the increase in speed is a function of the amount of practice. “There do not appear to be any cognitive limits on the speed with which a skill can be performed” (p. 236). The speed effect can be described by a function:

$$T = a * P^{-b} \quad (T\text{: time, P: practice})$$

Anderson described the case of a woman whose job was to roll cigars in a factory. Her speed at cigar making improved continuously over 10 years. This improvement through practice is of the same type as in general memory tasks. For instance, Anderson (1983) had subjects practice reading sentences like (A) for 25 days and looked at the effects of this practice on time to recognize a sentence:

(A) The sailor is in the park.

Figure 2.5 illustrates the results. Subjects' speed increased from 1.6 seconds to 0.7 seconds. Thus, retrieval time was cut by more than 50%. It is important to note, however, that the rate of improvement decreases with even further practice.

The theoretical framework of Anderson's considerations was provided by cognitive psychology. Cognitive psychology states that information used by human beings is cognitively represented. This means that a sentence like (A) is not only a sequence of signs measurable in bits—but for the sentence to be understood, the information has to fit into a certain cognitive representation of the world involving knowledge about sailors and parks. Thus, the idea of representation is central to cognitive psychology. Building chunks out of information is more than reducing informational complexity: Chunks must fit into the domain representation we acquire when becoming experts. "Thus one important dimension of growing expertise is the development of a set of new constructs for representing the key aspects of a problem" (Anderson, 1985, p. 256).

In his book *Cognitive Psychology and Its Implications*, Anderson provided a general explanation for the acquisition of skills. According to Anderson, becoming expert in a domain means: Declarative (textbook-like) knowledge changes into (automatic) procedural knowledge. Declarative knowledge is "knowledge about facts and things"—it is the knowing what. Procedural knowledge is "knowledge about how to perform various cognitive activities"—it is the knowing how. Typically, novices start by rehearsing the facts relevant to a skill. For instance, they learn the different chess pieces, the different open-

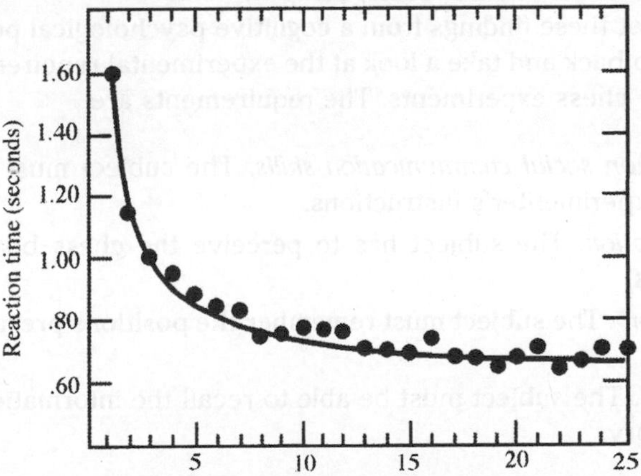

FIG. 2.5. Time to recognize a sentence as a function of the number of trials of practice. *Note.* From *Cognitive Psychology and Its Implications* (p. 146) by J. R. Anderson, 1985, San Francisco: Freeman. Copyright 1985 by Freeman. Reprinted by permission.

ings and stages of a game. This is declarative knowledge. In time, the facts to be learned do not need any further consideration. Pawns and bishops are drawn without deliberation on what rules might refer to them.

The procedural form of knowledge consists of procedures (i.e., if–then rules that can be automatically processed). For example, a chess procedure might be: "If the goal is to regain a lost Queen, then you have to cross the board with a Pawn." From Anderson's point of view, the speed of expert performance is due to the system of memory-based rulelike procedures. Each time a specific if condition is fulfilled, the then conclusion follows automatically. The transformation of more or less general declarative knowledge into specific procedural knowledge explains not only experts' time gain, but also the domain specificity of expert performance. The if condition of expert procedures is restricted to the scope of the domain experience an expert has gained.

The Dynamics of Categorical Perception

Cognitive psychology uses models of human cognition. Short-term memory is such a model for memory processes. Although the concept of short-term memory is, according to Anderson (1990), "much in dispute today," we can state some basic cognitive facts: "(a) we can maintain rapid and reliable access to but a few items, and (b) once unattended, those items rapidly decay" (p. 91). He added three more facts in connection with experts: (c) experts have rapid access to highly complex items, (d) these items are domain-specific, and (e) there is much practice needed to gain mastery—that is, to acquire the relevant domain-specific, complex items.

To interpret these findings from a cognitive psychological point of view, we have to go back and take a look at the experimental requirements of the deGroot type chess experiments. The requirements are:

(a) *Common social communication skills.* The subject must understand the experimenter's instructions.

(b) *Perception.* The subject has to perceive the chess board and the pieces.

(c) *Memory.* The subject must remember the positions presented on the board.

(d) *Recall.* The subject must be able to recall the information stored in memory.

Each step defines a necessary condition for the following steps. If a person lacks the necessary social communication skills, the experiment cannot be conducted. If for whatever reason the subject is unable to perceive the board, it is useless to proceed with the experiment. The same is true if the subject lacks memory, and so on.

Putting the common social communication skills aside, we see that perception is the first crucial cognitive process. Cognitive psychology stresses the fundamental insight that perceiving means reconstructing. The world we experience is a reconstruction of what might be reality outside. Cognition starts with perception in its very sense. Thus, it is not surprising that in chess—as in many other domains—expertise involves special perceptual skills. Masters see more than beginners or weaker players do—namely, they see more relevant things. In fact, they do not see a board with pieces on it, but a strategic position in a game. Thus, perceiving is the integration of memory with sensory data.

deGroot spoke of *dynamic perception*. Today this phenomenon is investigated in the context of *categorical perception* (Harnad, 1987). Categorization helps us and other organisms to master the complexity of the environment in which we live. Categorical perception means that perception not only provides the sensory data for higher cognitive processes, but also a model for categorization in general. Categorical perception can be understood as "an analog-to-digital transformation that recodes a continuous region of physical variation as a discrete, labeled equivalence class" (p. 4). For an interpretation, we may reconsider experiments by Tajfel in the 1960s.

Henri Tajfel, a social psychologist, used a simple experiment to show the effects of categorization. He had subjects judge the length of eight painted lines. The lines differed from each other by a constant ratio of 5% of length. The four shorter lines were labeled as Category A—the category of *short lines*. The longer ones were labeled as Category B—category of the *long lines*. As a consequence of this simple categorization, subjects overestimated the difference between the longest Category A line and the shortest Category B line. The Category B line belonging to the long lines was judged much longer than the Category A line that belonged to the short lines. Thus, variation among categories is overestimated at the expense of variation within each category.

TABLE 2.2
Categorization of Two Groups of Lines

Lines *Category*	*1* *A*	*2* *A*	*3* *A*	*4* *A*	*5* *B*	*6* *B*	*7* *B*	*8* *B*
Real length	16.2	17.0	17.9	18.8	19.7	20.7	21.7	22.8
Mean estimated length	16.0	17.3	18.1	19.3	21.1	22.3	23.6	25.3
Without categorization	16.4	17.3	18.2	19.3	20.3	21.5	22.6	24.2

Note. Data are from "Classification and Quantitative Judgment" by H. Tajfel and A. L. Wilkes, 1963, *British Journal of Psychology, 54*, Table 2. Copyright 1963 by The British Psychological Society. Reprinted by permission. The persons being presented with the categorization overestimate the long B lines, especially the difference in length between Line 4 (the longest of the short A lines) and Line 5 (the shortest of the long B lines).

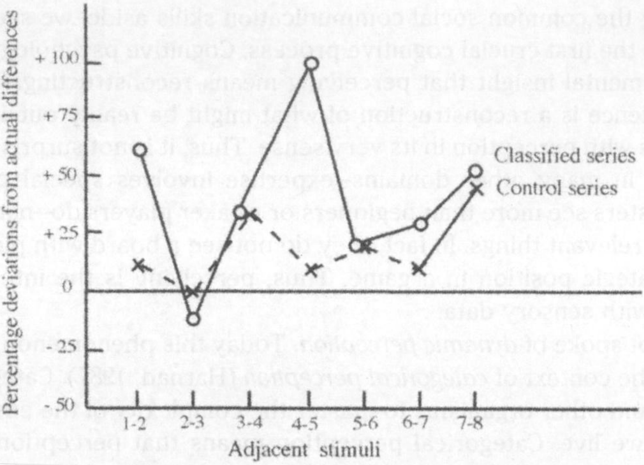

FIG. 2.6. Categorization. Comparison of actual and estimated differences between adjacent stimuli (lines). The difference is maximal for the stimuli 4 (the longest of the short A lines) and 5 (the shortest of the long B lines), which means at the border between short and long lines. In the absence of categorization (control series), the overestimation disappears. *Note.* From "Classification and Quantitative Judgment" by H. Tajfel and A. L. Wilkes, 1963, *British Journal of Psychology, 54*, Fig. 1. Copyright 1963 by British Psychological Society. Reprinted by permission.

Categorization speeds up cognition. Categorization helps us distinguish between the relevant and irrelevant. In chess, being able to see which of your pieces are endangered enables you to search for the right move in advance. Categorization is a simple but effective representation of the domain. It is prolonged domain-specific experience that enables experts to categorize the domain. This is the basis of expertise. As deGroot (1965) put it, experience is "the foundation of the superior achievements of the masters" (p. 329). In the terms of cognitive psychology, experienced categorization enhances the fast activation of effective procedures.

Another aspect of the development of expertise is deliberation. Anders K. Ericsson, who investigated the reality of the 10-years rule in several domains, especially in skills such as playing the violin, stressed the importance of feedback and training. Performance in almost every domain can be increased by feedback and training. Deliberate practice plays an important role in the acquisition of expert performance (Ericsson et al., 1993). If we consider categorization fundamental to expert performance, we have to say: Experts have to undergo a long period of perception schooling.

Experts in some domains outperform novices by the simple fact that they find the suitable categorization or cognitive representation of a problem. From this point of view, many important results of expert–novice research become obvious:

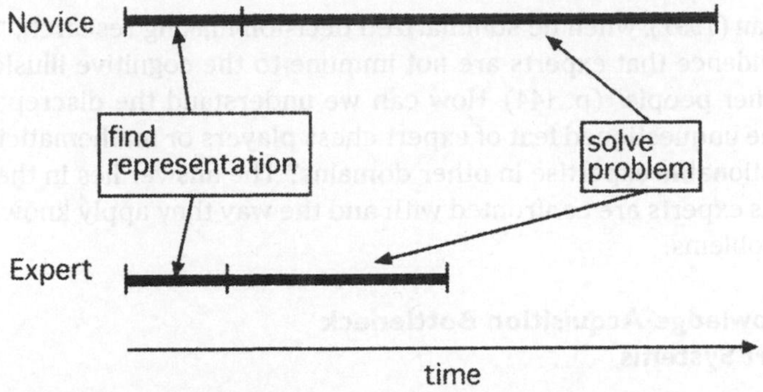

FIG. 2.7. Experts spend proportionally more time building up a basic representation of the problem situation before searching for a solution. *Note.* From *The Nature of Expertise* (p. 312) by M. T. H. Chi, R. Glaser, and M. J. Farr, 1988, Hillsdale, NJ: Lawrence Erlbaum Associates. Copyright 1988 by Lawrence Erlbaum Associates. Reprinted by permission.

- Experts spend proportionally more time building up a basic representation of the problem situation before searching for a solution (e.g., Chi, Glaser, & Rees, 1982; Lesgold et al., 1988).
- A schema or categorization with the high probability of being at least in the right problem space is invoked very rapidly (Lesgold, 1984).
- Expert problem solvers tend to work forward from the given information to the unknown (see e.g., Patel & Ramoni, 1997). In general, experts do not start with generating a set of hypotheses and test each of them subsequently. Rather, metaphorically spoken, they use the given information to "walk through" their domain knowledge.

Thus, expert categorization is a fast interaction between sensory data and long-term memory. Categorization aids cognitive economy. To summarize: Cognition means representation. Through their experience, experts acquire the ability for domain representation. Experts are fast due to their grasp of categorizations within the domain representation and the use of domain-specific procedures.

2.3 ECONOMIZING HUMAN EXPERIENCE: EXPERT SYSTEMS AND PROFESSIONS

The story so far on expert performance is somewhat too bright. Psychology has examined a variety of domains for expertise. Many experts such as physicians, stockbrokers, court judges, and others seem to err more often than we would expect when it comes to decision making. According to Daniel

Kahneman (1991), when he summarized decision-making research, "there is much evidence that experts are not immune to the cognitive illusions that affect other people" (p. 144). How can we understand the discrepancy between the unquestioned feat of expert chess players or mathematicians and the questionable expertise in other domains? The answer lies in the kind of problems experts are confronted with and the way they apply knowledge to these problems.

The Knowledge-Acquisition Bottleneck in Expert Systems

Research on expert decision making commenced with analyses of the validity and reliability of expert judgments. *Validity* means accuracy; a valid judgment is a true one. Reliability is often measured as repeatability: A decision is reliable if the decision repeated under the same conditions results in the same judgment. In many domains, the validity and reliability of expert judgment can be doubted. The first known study on experts reported that grain rated highest by corn judges did not always produce the highest crop yields; there was also considerable variability among corn judges (Hughes, 1917; Shanteau & Stewart, 1992). A more recent study found that grain judges misjudged nearly one third of wheat samples. Moreover, judged a second time, over one third of the samples were graded differently (Trumbo, Adams, Milner, & Schipper, 1962).

Expert judgment has been thoroughly investigated in clinical settings and in physicians. In the 1950s, a controversy started on clinical versus statistical judgment in clinical psychology (Meehl, 1954). The clinical judgment is made by an expert (e.g., a psychologist or physician). The statistical judgment is derived by a computer. Clinical psychologists argued that one could not make proper use of a psychometric test (e.g., Minnesota Multiphasic Personality Inventory [MMPI]) without extensive training and experience. However, a series of experiments showed that:

- in several cases, the judgments of expert clinicians were no better than those of graduate students (Oskamp, 1967); and
- statistical diagnoses—regression models—outperformed clinicians' judgments (Goldberg, 1969, 1970).

Research on decision making by medical doctors brought similar findings. In the clinical context, the overall predictive ability of clinical judgment was found to be poor, sometimes zero (see Einhorn & Hogarth, 1978). In contrast, self-confidence in judgments made by clinicians was found to increase when more information was available, but without any corresponding increase in judgmental accuracy (Oskamp, 1965)—note that a similar effect had been found in other fields of expertise such as corn judges (Shanteau & Stewart, 1992). Einhorn and Hogarth (1978), two decision-making researchers, spoke of the illusion of validity: Even experts seem to be overconfident of their judgments. A wide-ranging research project on medical problem solving, "The Medical Inquiry Project" at the Michigan State University from 1969 to 1973, could not dispel doubts on clinical decision making (Elstein, Shulman, & Sprafka, 1978). In contradiction to standards of rational decision making, medical doctors looking for a diagnosis generated a leading hypothesis very early. Also clinicians were found to have a distinctly limited capacity when it came to simultaneously considering multiple hypotheses. Regarding decision making, experts do not seem to differ from normal people (Elstein et al., 1978).

The research on expert decision making stands in a line of science different to cognitive psychology because it is not the cognitive process but the resulting decision that is examined. Rational criteria can be applied to decision making, thus, decision-making research compares expert decisions to ideal decisions derived from rational decision-making models.

Decision-making research is rooted in the classics of utility theory and rational choice, such as von Neumann and Morgenstern's (1944) *Theory of Games and Economic Behavior*. The concept of decision making rests on the idea of rational choice: A decision consists of choosing between alternatives. For instance, a doctor has to choose the appropriate therapy for a patient. If you have to decide to undergo medical surgery, your decision can be traced back using a decision tree that shows the decision alternatives—therapy or no therapy—and the possible outcomes that range from death to recovery. To apply decision theory, you have to weigh the alternatives by combining the probability of the alternative with the utility of the outcome. Then if the expected utility of the therapy alternative weighs more than the expected utility of the no-therapy alternative, by ratio the decision must be to have therapy (see Fig. 2.8).

We can find the ideal of rational decision making implemented in the research efforts of artificial intelligence (AI)—the most prominent examples being rule-based expert systems. The construction of expert systems seemed to overcome the fallibility of expert judgments through rational decisions. A rule-based expert system is basically a computer or a computer program; its structure is composed of two elements: (a) the shell, and (b) the knowledge base.

The shell is an inference mechanism. The knowledge base contains the knowledge specific to the discipline. This knowledge is stored in the shape of facts and rules—corresponding to the difference between declarative and procedural knowledge. The rules have the if–then form like a procedure. The most widely known medical expert system development was Mycin. Starting in 1973, Mycin was developed at Stanford University. It was in-

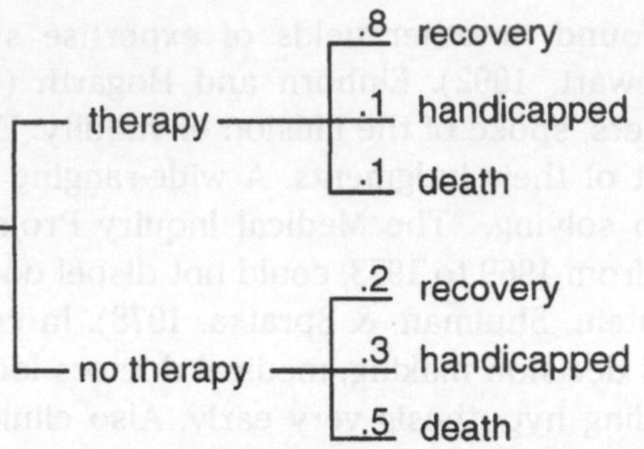

FIG. 2.8. Simplified decision tree for the decision to undergo medical therapy. The probabilities of the possible outcomes are written on the lines. To weigh the alternatives, the possibilities of the outcomes are to be multiplied with their utilities. The utility (u) of death, for instance, might be zero. If we feel the utility of recovery counts about 10 times more than the state of being handicapped for the rest of your life, we can set:

u(death) = 0
u(recovery) = 10
u(handicapped) = 1

The expected utility of the therapy alternative is 8.1 (= .8 × 10 + .1 × 1 + .1 × 0) and outweighs the expected utility of the no-therapy alternative, which is 2.3 (= .2 × 10 + .3 × 1 + .5 × 0). Thus, we have to decide in favor of the therapy alternative.

tended for medicative therapy of certain bacterial infections. To develop Mycin, bacterial infections had to be described as a decision problem:

> Selection of therapy is a four-part decision process. First, the physician must decide whether or not the patient has a significant infection requiring treatment. If there is significant disease, the organism must be identified or the range of possible identities must be inferred. The third step is to select a set of drugs that may be appropriate. Finally, the most appropriate drug or combination of drugs must be selected from the list of possibilities. (Buchanan & Shortliffe, 1984, p. 14)

The Mycin rules are weighted according to a kind of transition probability, so-called *certainty factors*, reflecting the certainty of a judgment. A Mycin rule reads as follows:

(B) IF: (1) The stain of the organism is gram positive, and
(2) The morphology of the organism is coccus, and
(3) The growth confirmation of the organism is chains,
THEN: There is suggestive evidence (.7) that the identity of the organism is streptococcus.

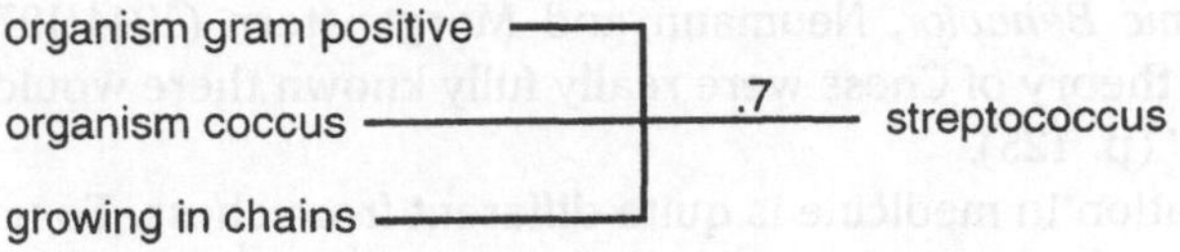

FIG. 2.9. Simple formal description of the Mycin rule (B).

The authors commented: "This rule reflects our collaborating expert's belief that gram-positive cocci growing in chains are apt to be streptococci. When asked to weight his belief in this conclusion, he indicated 70% belief that the conclusion was valid" (Shortliffe & Buchanan, 1984, p. 238).

Expert systems intended to provide expert knowledge in a conclusive form. For this to occur, however, an intensive collaboration and interrogation of experts is necessary. A central problem is the knowledge-acquisition bottleneck. The problem is how to get the necessary knowledge into the system. The problem starts with experts being mostly unaware of what they know (Nisbett & DeCamp Wilson, 1977). The kind of knowledge that expert system engineers need from experts is declarative. As already described, however, as an effect of the routinization of expert work, experts' knowledge becomes more and more procedural. Not only do domain experts lack the ability for reliable introspection about the highly skilled knowledge they use during proficient problem solving, but they are also often reluctant to weigh the certainty of knowledge using quantitative measures. What is the evidence for the certainty of streptococcus of .7 and not of .6 or .685? The knowledge-acquisition bottleneck problem becomes even worse once a considerable amount of valid information has been implemented. The problem is how to update the knowledge base when new categories of knowledge—new bacterial infections—appear. The problem refers to the failure of an expert system to offer appropriate advice on classes of cases that the expert system developers have not considered at the time they created the system. "The difficulty may lie not so much with the bottle as with the content" (Musen & van der Lei, 1989, p. 29).

Although the experimental evaluation of Mycin was successful (Yu et al., 1984), Mycin was never used in a clinical setting. As to medicine and related fields, the fate of Mycin is not an exception (see Shortliffe, 1989). The problems with the expert system concept affect the scope and functioning of our knowledge (see Mieg, 1993). Expert systems of the Mycin type cannot work without the unspoken assumption of a closed world. For instance, a closed world can be found with chess problems. There is a finite number of pieces with well-defined functions. The chess board has a limited number of fields. As a consequence, at each stage of a chess game, there is a limited number of possible moves. The problem of the best next move can be studied and solved using decision trees like the one in Fig. 2.8. In their *Theory of Games and Economic Behavior*, Neumann and Morgenstern (1944/1972) asserted that, "if the theory of Chess were really fully known there would be nothing left to play" (p. 125).

The situation in medicine is quite different from chess. To use the chess metaphor: In medicine, it is still possible that new pieces appear with unknown functions, for instance, new viruses; and even the playing ground, the chess board, does not have a constant size. There is no final conclusion on whether biological models or psychological therapies have to be applied to many clients with psychiatricly relevant problems (in short, whether to prescribe pills or psychotherapy).

The knowledge-acquisition bottleneck problem reveals some general characteristics of expert knowledge: (a) Expert knowledge is domain-specific, and (b) the relevance structure of the domain has once again to be determined by experts, thus leaving no way of using expert knowledge without using experts. Let us discuss both points in more detail.

Expert knowledge is *domain-specific.* The social psychologist James Shanteau reexamined studies on expert decision performance in several domains, decision performance being measured by the reliability and validity of judgment. Experts with good decision performance can provide good predictions. Shanteau distinguished two sorts of domains differentiated by the decision performance found in expert judgment (see Table 2.3). The findings for physicians diverged: For example, radiologists show good decision performance, psychiatrists a poor one.

TABLE 2.3
Domains With Good or Poor Expert Decision Performance

Good Performance	*Poor Performance*
Weather forecasters	Clinical psychologists
Livestock judges	Psychiatrists
Astronomers	Astrologers
Test pilots	Student admissions
Soil judges	Court judges
Chess masters	Behavioral researchers
Physicists	Counselors
Mathematicians	Personnel selectors
Accountants	Parole officers
Grain inspectors	Polygraph (lie detector) judges
Photo interpreters	Intelligence analysts
Insurance analysts	Stockbrokers
Nurses	Nurses
Physicians	Physicians
Auditors	Auditors

Note. From "Competence in Experts: The Role of Task Characteristics" by J. Shanteau, 1992a, *Organizational Behavior and Human Decision Processes, 53*, p. 258. Copyright 1992 by Academic Press. Reprinted by permission.

Shanteau (1992b) presented several explanations for this dichotomy in expert judgment performance:

(a) First, "domains where good performance has been observed involve decisions about objects or things," as shown in the left-hand side of Table 2.3.

(b) Another reason might be that "observed predictability is different for the two sides: human behavior is inherently less predictable than physical stimuli," the right-hand side of Table 2.3 containing the expertise related to human behavior.

(c) Finally: "With domains on the left-hand side, there are more chances to learn from past decisions." (p. 15)

Shanteau (1992b) concluded that "the skills and abilities that emerge (or don't emerge) in experts depend on the situation in which they work" (p. 14). In chess, with the closed world of the board, the task is clear-cut. Experience and training will not be confronted with new phenomena—a situation different to that of psychiatry or personnel selection.

However the differences in predictive power might be caused, knowledge in the domains in both columns of Table 2.3 is structured by experts. Also, all the knowledge implemented in an expert system is provided and defined by domain experts. This is true even for the criteria used to validate expert decision performance. For instance, to evaluate a diagnosis of psychosis, other psychiatricly relevant categories have to be used, such as hallucinations, bizarre behavior, inappropriate affect, and so on. In most cases, validation is obtained through the judgment of other domain experts. Thus, any judgment on expert decisions has to be based on the knowledge of domain experts. Computers are more reliable when it comes to using diagnostic criteria established by experts. The use of computers guarantees that one and the same diagnosis is reliably repeated. Therefore, it is hardly surprising when computers outperform human experts, attaining more correct diagnoses.

Expert knowledge is a response to uncertainty in a domain relevant to society—from mental health to weather forecasting. Therefore, in some domains, it is better to have poor expert predictions than to have no predictions at all. Expert knowledge structures the problem spaces. It offers criteria needed to understand domain-specific problems. Experts define relevant properties, possible diseases as well as possible causes, and possible remedies. To decide what is relevant is the core problem of the knowledge-acquisition bottleneck of expert systems. What behavior can be called bizarre? What affect is inappropriate? What are hallucinations—and not simply day dreams or visual illusions? Thus, experts de facto decide on relevance. For the time being, we and the expert systems have to rely on the prestructured concepts provided by domain experts. In contrast to human experts, expert systems cannot really comprehend the scope of relevant problems: A naive user of Mycin would have no way to ascertain the scope of Mycin's ignorance (Musen & Van der Lei, 1989). It is not so much a question of implementing enough knowledge; the very bottleneck of expert system-knowledge acquisition is the problem of knowing the relevant. Hoffman, Feltovich, and Ford (1997) summarized the history of expert systems: "There has developed a consensus that expert systems are not emulations of expertise, but tools to support experts as they go about their familiar tasks" (p. 550).

The Power of Abstraction

There is only a limited use of rational, rule-based expert systems. Expertise is about knowledge, not rational decision making. Expert knowledge is specialized; otherwise there would be no special demand for it. However, as seen in the expert–system–bottleneck problem, expert knowledge has a fuzzy border to commonsense knowledge that consists basically of the interface problem of communicating whether a problem is relevant to a domain of knowledge. Expert knowledge is not, as one might think, a clear-cut part of the knowledgeable world.

Expert knowledge is knowledge in use; this means that there is a demand for expert judgment. Even if expert judgments are not rational in comparison to some decision-making standards, they are generally considered to be the best judgments available. When we said that expert knowledge was a response to uncertainty in a domain relevant to society and that expert knowledge structured problem spaces, we left the field of cognitive psychology. In society, expert knowledge is administered and provided by occupations and organizations. The chief occupational form of expert labor is a profession.

Andrew Abbott's (1988) *The System of Professions* described how societies structure expertise. Professions such as medicine or law administer expert knowledge. Abbott stated that professions compete for social acknowledgment in solving special problems—for instance, curing illnesses or formulating contracts. Each profession claims to have the appropriate knowledge. Knowledge is "the currency of competition" (p. 102).

The link between a profession and a task—Abbott called it *jurisdiction*—can change. The main mechanism of jurisdiction shift is abstraction. Abstraction, according to Abbott (1988), takes on two forms. First, abstraction can be reduction in the sense of detaching from specific content, that is "abstract which refers to many subjects interchangeably" (p. 102). For instance, psychology claims alcoholism to be a personality disorder, thus falling into the domain of psychotherapy. By abstraction, alcoholism may also be seen as a medical problem or a problem of public administration. The ultimate abstraction is the concept of expert systems and the claim of AI to reduce all forms of occupational work to one programlike form.

Abstraction in the second sense is formalization, saying "that knowledge is abstract that elaborates its subjects in many layers of increasingly formal discourse" (Abbott, 1988, p. 102). Formalization means that a profession provides routines and formulas for problem solving. Formalization strengthens the jurisdiction of a profession. "No one tries to explain particle interactions without mastering the abstract knowledge of physics. More practically, no one offers insurance companies advice on underwriting without having mastered actuarial theory" (p. 103).

Extreme abstraction can weaken a jurisdiction. Abbott (1988) cited the area of business management. Despite numerous efforts, no successful exclusive jurisdiction has been established in this field; business management lacks effective tools and treatments. The problem with business management is "the tenuous connection between the various abstractions applied to the area and the actual work of managers" (p. 103). As a result, there are numerous professions claiming jurisdiction to business management, such as economics, administration, psychology, and others.

Abstraction means abstraction of knowledge. The main source of abstract knowledge is science. Academic or formalized knowledge is held to be characteristic to professions (e.g., Parsons, 1968). Sociologist Everett C. Hughes (1965) spoke of detachment in professional work: "having in a particular case no personal interest such as would influence one's action or advice, while being deeply interested in all cases of the kind" (p. 6). Therefore, professional work has to maintain an optimum level of abstraction (Abbott, 1988, p. 105). Professions make believe that abstract academic knowledge is congruent with practical professional knowledge (p. 54). In reality, professional knowledge is related to work and is shaped by interprofessional competition. Thus, it would be difficult to understand a doctor's day-to-day work as a realization of abstract medical knowledge. Instead, the doctor's professional work is clearly distinguishable from the work of a nurse or the business of a cleric or spiritual healer.

According to Abbott, professions have to be seen as parts of a system. They are dependent on each other, linked by historical chains as well as actual competition. The system of professions structures expert labor—and expert knowledge—within society and covers the relevant problem areas. In a case study, Abbott described the appearance of personal problems as a consequence of the social and industrial developments in the 19th century. In a rough summary: At the beginning of the 20th century, three professions—the clerical profession, neurologists, and psychiatrists—struggled to gain control of the personal problem market. The clergy lost because of their extreme abstraction of personal problems as salvation problems. In contrast, neurology offered a sophisticated diagnostic systems, but lacked special therapies that were not part of general medicine. Only psychiatry—at that time—could provide for effective therapy offering psychotherapy on the basis of psychoanalysis.

Knowledge abstraction faces two constraints: (a) the kind of problems the knowledge is applied to, and (b) the market of expert labor. Thus, validity and reliability of expert judgments strongly depend on these two factors. Most of the knowledge domains in Shanteau's list (Table 2.3), where experts display poor decision performance, are related to the assessment of human behavior—a domain claimed by many professions. As to the assessment of human behavior, there might be a vicious, competitive connection between the assessment and behavior assessed. In the case of personnel selection, for instance, the person assessed might know the selection criteria and try to fake or imitate the behavior requested, or at least dispute the validity of the selection criteria. Thus, the assessment of human behavior can force a reaction by the persons assessed and change the conditions for further assessments. As a consequence, predicting human behavior is a difficult, reflexive task. A similar argument might be true for stockbroking, but obviously not for chess games.

To summarize: Experts are not—as far as generalizations are possible—better decision makers than are normal persons. Differences in performance are due to differences in knowledge and task domains. As far as knowledge in our societies is administered by professions, different knowledge systems compete.

2.4 ECONOMIZING COGNITION: COLLECTIVE RATIONALITY

In the preceding paragraphs, we used the terms *economy* and *economizing* without clear definitions. What do we mean by economy? Following the sociologist Max Weber (1979), we can say:

> We shall speak of economic action only if the satisfaction of a need depends, in the actor's judgment, upon relatively *scarce* resources and a *limited* number of possible actions, and if this state of affairs evokes specific reactions. Decisive for such rational action is, of course, the fact that this scarcity is *subjectively* presumed and that action is oriented to it. (p. 339)

On the one hand, there are needs and a demand; on the other hand, there are limited means and a limited supply, and this discrepancy causes specific transactions. The same also applies to expert knowledge. There is a specific need and demand, some sort of uncertainty, caused by limited knowledge resulting in a demand for expertise.

Gary S. Becker (1993) spoke of *human capital* that is created by investments—education, training, medical care, and so on (p. 16). Experts can be regarded as a form of human capital. Human capital is capital in the sense that it can be invested in industries to raise productivity. The long periods of persisting growth in per capita income in the United States, Japan, and some European countries presumably are, as Becker stated, due to "the expansion of scientific and technical knowledge that raises the productivity of labor and other inputs in production" (p. 24). If this is true, we can assume that the contribution of human experts to the increase in productivity is crucial and not limited to engineering the technical basis of industries.

What is new about understanding experts as human capital? Isn't that simply another way of speaking of the division of labor? The notion of the division of labor could invoke ideas of a preestablished categorization of possible occupations. From this point of view, experts are specialists for specific problems. In this case, it would be best to educate everyone at the place he or she will go to work for the rest of his or her life. However, this would underestimate the dynamics of productive knowledge in our societies.

The discussion on expert decision making reveals the importance of expert knowledge. However, what are experts supposed to supply: knowledge, problem solving, decision making? This question accompanies us throughout the whole book. To arrive at more precision in determining the labor of experts, we have to discuss rationality. In one sense or another, expert labor, be it expert advice or expert problem solving, is thought to be rational.

Satisficing

Herbert A. Simon (1972), economist, psychologist, and one of the founders of the research program on AI, described *limits of rationality* that affect human beings and other information-processing living organisms. Simon called theories on human behavior that reflect these limitations "theories of bounded rationality." Theories of bounded rationality may incorporate:

(a) *Risk and uncertainty*. The consequences of decisions or some behavior may not be certain. This is the situation a physician has to live with.

(b) *Incomplete information about alternatives*. For instance, a physician, given a client with some symptoms, may not know all possible therapies in detail, let alone all side effects.

(c) *Complexity*. Referring to the economic situation of calculating costs and returns, Simon says that rationality may be "bounded by assuming complexity in the cost function or environmental constraints so great as to prevent the actor from calculating the best course of action." (p. 164)

The limits of rationality are not simply defects of cognition. They have an ecological function: the adaptation to complex real-world environments. Simon (1990) asserted, "Human rational behavior is shaped by scissors whose two blades are the structure of task environment and the computational capabilities of the actor" (p. 7). The idea of an adaptive mind is widely recognized. Anderson (1990) called the adaptive function a "General Principle of Rationality: The cognitive system operates at all times to optimize the adaptation of the behavior of the organism" (p. 28). The adaptive behavior of an organism falls far short of the economic principle of maximizing. Being thirsty, an organism would not try to get the drink that is best in every aspect: flavor, vitamins, quantity—but to get the first one. Simon (1956) used the word *satisfice* rather than *maximize* or *optimize*. According to the satisficing principle, we do not strive for the best of all possible actions; instead of calculating all possible outcomes, we decide for the first alternative that satisfies our aspiration level. "Organisms adapt well enough to 'satisfice'; they do not, in general, 'optimize' " (1956, p. 129). *Satisfice* is a word of Scottish origin—a word used for procedures that meet the constraints of limited time, knowledge, or computational capacity.

Generally speaking, in every process of decision making, the principle of rationality is held to be applicable. Simon saw no difference between human individuals and organizations. He claimed that a theory that explains individual cognition also covers the neural organization of the brain and therefore has "a significant organizational component," whereas an organizational theory that is capable of treating an organization as a monolith is by virtue also a theory that explains the behavior of individuals (Simon, 1972, p. 161).

Human thinking, seen under its adaptive function, normally serves some kind of problem solving. The problems might consist of satisfying basic needs such as finding food or, more common human activities, finding a good job, opening beer cans, or solving puzzles. According to Simon, cognitive problem solving under the condition of bounded rationality is characterized by the use of heuristics. Heuristics are rules of thumb. They help cut down on the variety of possible decision branches in the decision tree. In chess, for instance, experienced chess players, even in difficult positions, normally do not examine more than 100 continuations. Such heuristics of problem solving are:

(a) *Means–end analysis*. 1. "The present situation is compared with the desired situation (problem goal), and one or more differences between them noticed" (1992, p. 67). Simon's example: I have a board five feet long; I want a two-foot length; there is a difference in length. 2. Memory is searched for procedures to overcome the difference (e.g., sawing). 3. If the procedure selected doesn't solve the problem (the board is too thick to be sawed), other procedures are searched for (buy another board of appropriate length).

(b) *Planning*. "Planning consists in omitting some of the detail of the actual problem by abstracting its essential features, solving the simplified problem, then using the simplified problem as a guide, or plan, for solution of the full problem" (Simon, 1992, p. 68). Planning in this sense is routinized in constructing: If a huge tower or a racing car is to be constructed, you first build a model to test the qualities of the intended product. The test might cause some changes to the model. Then the tower or car is constructed according to the model.

(c) *Factoring a problem into subproblems*. This enables you to solve only minor subproblems. For instance, running a hotel has to be cut down into subproblems: the building, the personnel, the bookkeeping, and so on.

However, heuristics are procedures that do not guarantee that a solution will be found. The tower built in accordance to its model might collapse because of wind or unpredicted static defects. Normally heuristics do not stand for the ideal solution. Instead they are good enough and satisficing—they reach the aspiration level the problem solver has.

Digression: Aspects of Rationality

There is much dispute about the so-called *homo economicus*. The discussion refers to humans as subjects of economic processes—maybe consumers, traders, or brokers. The questions discussed are: How rational are economic decision makers? In what way are they rational? We do not want to review or repeat the concepts. We only want to elicit some characteristics of the kinds of rationality under discussion. We discuss four points:

1. *Means–goals adequacy*. There is a common understanding of rational behavior. Simon (1976) called it substantive rationality: "Behavior is substantively rational when it is appropriate to the achievement of given goals within the limits imposed by given conditions and constraints" (p. 129). This kind of rationality urges you to calculate the means needed to achieve a goal. If you want to read a newspaper, you do not need to buy a publishing house. You certainly will buy the newspaper on the street, but are not willing to pay more than the normal, appropriate price.

Given a goal, the rational achievement of the goal is a problem of allocating resources, such as capital and workforce, thus maximizing utility. The assumption of substantive rationality is the starting point of classical economics.

2. *Procedurality*. Rationality is often interpreted as some sort of calculated procedure. Simon (1976) defined *procedural rationality*: "Behavior is procedurally rational when it is the outcome of appropriate deliberation" (p. 131). Procedural rationality can also be characterized by the *Newell principle*: "If an agent has knowledge that one of its actions will lead to one of its goals, then the agent will select that action" (Newell, 1982, p. 102; see also Anderson 1990, p. 15).

In the model of rational decision making, you have to calculate probabilities and values to execute the best alternative (e.g., to find the best therapy for the diagnosed disease). Once the best alternative has been found, it should—by rationality—be put into reality.

3. What about logic? The human limitation of rationality—in the sense so far discussed—is not necessarily due to limits in logic, but in computation. There is some dispute regarding the average human computational capability as it relates to mathematical models of probability that are necessary for modeling rational decision making. Our computational limits can be shown much more easily when it comes to exponential functions like 2^n. There are two classic examples that demonstrate our poor understanding of exponential processes:

- The first example is related to the mythical origins of chess. The king who had ordered the invention of chess, the royal game, was so pleased by the result that he offered the inventor any reward the inventor wished and the king could grant. The inventor's simple wish was to get one grain of rice for the first field of the chess board, two grains for the second, four grains for the fourth, and so on. The king accepted the price, which he thought to be a quite humble wish. But the 2^{63} grains of rice the king would have needed to pay the 64 fields of the chess board were in fact more rice than had ever grown on Earth.
- The second example has to do with simple newspapers. Normally newspapers have an uncomfortable size. If you want to carry them, you usually fold them once or twice. The question: How often, if you could, do you have to go on folding a single sheet of a newspaper, laying one half of the sheet on the other half, so that the resulting package reaches the height of the distance between Earth and the moon? The answer is about 43 times. After folding the newspaper about 82 times, you would have already passed the limits of our galaxy.

The growth of exponential functions is outside of our everyday estimation capabilities. This constrains our everyday thinking about biotic processes that sometimes show exponential growth characteristics. However, does this prove that we cannot master logic?

One of the psychological experimental procedures that reveals the difference between calculation and logic was invented by Peter Wason (see

Wason & Johnson-Laird, 1972). The experimenter lays out four cards in front of you, displaying the following symbols.

FIG. 2.10. Card experiment.

Each card has a number on one side and a letter on the other side. There is a hypothesis concerning these cards, saying that:

(H) If a card has a vowel on one side then it has an even number on the other side.

The task is as follows: Which cards do you have to turn in order to prove the hypothesis? You should not turn more cards than necessary.

The task is not easier to solve than to understand. Many subjects are tempted to turn the card with the number 4, but there is no need to do that: Even if the card with the 4 on the one side has a consonant on its other side, Hypothesis H is not falsified; H does not say a word about cards with consonants. Instead of now discussing all the possibilities of checking cards, we turn to an analogous task. Imagine that you are cashier in a bank. During a day, four customers have changed checks for cash. There is a Rule R, saying:

(R) Every check for over 1 million dollars that is cashed has to be signed and approved by a director of the bank.

The checks show the amount on one side, and have, if necessary, to be signed on the back. You have four checks in front of you, two of them showing their amount and two of them their back; one check displays the signature of the director.

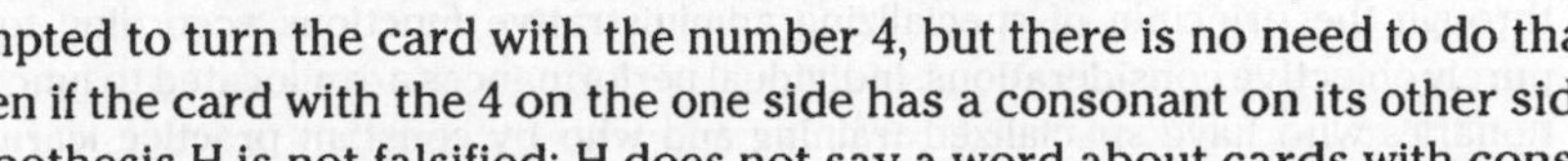

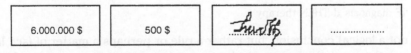

FIG. 2.11. Check version of the card experiment.

Which checks have to be turned in Fig. 2.11 to check if everything is okay? You will easily find that you only have to turn the first and last checks. You will also find that the check and card problems have the same structure, the solutions are equivalent. Versions of the card problem have generally proved to be more easily solved in realistic settings (Johnson-Laird, 1983, pp. 31–32). The card problem has a textbook-like form and it provokes some kind of calculation. However, if we find a naturalistic cognitive representation of a problem, it can be solved without computing—but by having the necessary experience that allows you to see the solution.

4. *Objectivity*. From a sociological point of view, rationality can also be understood as some form of objectivity. Rationality in this sense especially refers to public administration and means impersonal application of bureaucratic rules. Max Weber used the term *rational* to characterize a specific type of power and authority. Rational authority is ruled by law. Every subject is treated in accordance to the same law without deflection of the personal goals of the functionaries. This is the core of the ideal type of rational legal authorities:

> There is the principle of fixed and official jurisdictional areas, which are generally ordered by rules, that is, by laws or administrative regulations. (Weber, 1946, p. 196)
>
> Bureaucratization offers above all the optimum possibility for carrying through the principle of specializing administrative functions according to purely objective considerations. Individual performances are allocated to functionaries who have specialized training and who by constant practice learn more and more. The "objective" discharge of business primarily means a discharge of business according to calculable rules and "without regard for persons." (p. 215)

Objectivity as an aspect of administrative rationality does not only mean nonsubjectivity or impersonality, but also oriented on the matter of fact. In this sense, rational administration does not treat persons but problems of a certain kind: tax payers, consumers, and so forth. Many everyday activities in our societies are ruled according to this model of impersonal, fact-driven administration: taxation, medical care, schooling, and so forth. Weber (1946) said: "Rational bureaucratic administration is power by knowledge" (p. 129). As to bureaucratic rationality, the logic does not go with procedures, but with subsumption. In general, a procedure like

IF diagnosis d THEN therapy t

is not a logical conclusion, but rather a rule or perhaps a matter of fact. Instead, the purely logical conclusion comes from subsuming a case under a category or from instantializing the if condition:

Given:

1. For every person having diagnosis d, therapy t has to be applied.
2. John has diagnosis d.

We can conclude:

Therapy t has to be applied to John.

This logical procedure is well known as a *syllogism*. Thus, the logical core of the bureaucratic (and procedural) rationality has to do with subsuming someone or something under the application condition or to identify someone or something as a case of a defined species. This is the case in diagnoses as well as in determining someone's taxation class or deciding on who is subject to social programs.

The Economical Aspect of Expertise

Bureaucracy is one form of organizing the transactions of specialized (expert) knowledge; there are functionaries within the bureaucratic organization that have jurisdiction over special problems. The divisional organization of some industrial enterprises has this bureaucratic structure, installing for example a division for research and development, a division for cost accounting, a division for marketing, and so forth. Weber stressed the parallelism between bureaucratic and economical processes; he concluded his paragraph on bureaucratic objectivity by stating that, " 'Without regard for persons' is also the watchword of the 'market' and, in general, of all pursuits of naked economic interests" (1946, p. 215).

Professions–physicians, psychologists, lawyers–stand for another kind of organization of expert knowledge transactions. The professions claim jurisdiction for special problems, competing with one another through their knowledge. Professions as well as the sciences play an important role in cognitive economics.

We do not intend to create a new type of economic theorizing through the concept of cognitive economics. *Cognitive economics* means above all:

(a) The economic objects–the traded goods–are cognitive entities: knowledge and rationality. There is a demand and a supply for these goods, causing economic transactions.
(b) The transactions of these cognitive goods organize the collective cognition process that economizes individual cognition capabilities and human capital.

By using experts, we overcome the individual limitations of cognition. Like the distribution of commodities by the market, knowledge is distributed by using experts. We do not know if the knowledge distributed is true, but we can assume that the knowledge is the most competitive–among professions and other knowledge sources—and that the knowledge assembled is much more than a single person ever could master.

From an economic point of view, the decisive point for using experts is the knowledge-by-time ratio: The expert offers knowledge in a shorter time than it would normally take us. Rather than investing time in experiencing a special part of the knowledgeable world, we can ask an expert. Using experts is rational because it is a time-efficient use of knowledge. As we see, when discussing the role of experts, the profit results from using persons—experts—and not from the use of knowledge as such.

We speak of cognitive economics because the knowledge market is constrained by the cognitive peculiarities of human decision making. Human experience allows for fast individual recognition within a specific frame of the knowledgeable world. Experts are the human capital invested in this knowledge market. In the next chapters, we see how the rational use of knowledge through experts structures collective cognition and the knowledge market.

CHAPTER

3

Essentials of Experts-in-Contexts: "The Expert"-Interaction

When discussing experts and their roles, the basic idea is to understand *expert* as a form of interaction rather than as a person. Thus, almost anyone can–under certain circumstances–act as an expert. We see, even if there is sometimes a mystical note attached to experts, that the interaction involved in consulting an expert or, respectively, being consulted as an expert is based on a simple fact: There is somebody who seems to have knowledge that someone else is in need of.

This chapter develops the minimal explication of expert or the expert's role we have been looking for. As we have already said, we do not aim at a full description of all possible aspects of that role. Instead, we come to three assertions that might contradict certain more or less implicit scientific assumptions on experts. The assertions are:

(a) *The expert* refers more to a form of interaction than to a person,
(b) There are nonprofessional experts, and
(c) The core of the expert's role consists of providing experience-based knowledge that we could attain ourselves if we had enough time to make the necessary experience.

We discuss the aspect of interaction and define *the expert* as a social form of interaction (chap. 3.1). Then we look at the conditions and constraints of attributing someone as an expert (chap. 3.2). Finally, we try to determine what are, indeed, the reasons for which we consult an expert, particularly the role of the time gain in using experts as compressed experiences (chap. 3.3).

3.1 INTERACTION: "THE EXPERT" IS A SOCIAL FORM

Living in a Western society, we might come into the following situations:

- asking someone on the street for directions to the station
- consulting a physician
- speaking in court as an appointed expert on asbestos
- instructing a child how to lace his or her shoes
- watching a broadcast discussion with scientists on safety regimes for nuclear power plants

These situations differ in various aspects: the persons and consequences involved, the frequency and probability of the event, and its general social significance. Even the perspective differs: In some of the situations, we are asking somebody; in others, we are answering in some way. In whatever way the situations may differ, they share their form: Somebody explains a matter (what, how, and/or why) to someone else. In the following, we call this form "The expert"-interaction or, simply, "The expert."

The Form "The Expert." As Georg Simmel (1908/1971), a sociologist at the beginning of the 20th century, remarked, we can always differentiate between the contents of a social situation and its form:

> Any social phenomenon is composed of two elements which in reality are inseparable: on the one hand, an interest, a purpose, or a motive; on the other, a form or a mode of interaction among individuals through which, or in the shape of which, that content attains social reality. (p. 24)

In the examples cited earlier, we have a common form, but the interests involved can vary considerably. Providing explanation need not be the main intent of the person asked: The person on the street may be in a hurry and unwilling to stop; the doctor's main purpose might be to keep his or her practice running; the appointed expert's dominant motivation might be not to say anything wrong; the mother instructing her child might not want to have to lace her daughter's shoes herself; and the scientists may want to present themselves in the discussion as favorable as possible. Nevertheless, all of them provide explanations for someone else.

Explaining a matter to someone else is a form of interaction. Simmel emphasized that only through their interactions do persons form a society or, at least, participate in associations and social groups such as families, towns, or corporations. In social groups, too, we find common forms:

> [S]ocial groups which are the most diverse imaginable in purpose and general significance, may nevertheless show identical forms of behavior toward one another on the part of their individual members. We find superiority and subordination, competition, division of labor, formation of parties, representation, inner solidarity coupled with exclusiveness toward the outside, and innumerable similar features in the state, in a religious community, in band of conspirators, in an economic association, in an art school, in the family. However diverse the interests are that give rise to these sociations, the *forms* in which the interests are realized may yet be identical. (p. 22)

Let us return to experts. Consulting an expert is a form of social interaction; it also has this form of "somebody provides an explanation to someone else who has asked for it;" this is what we call "The expert"-interaction. Crucial to interaction—we are still following Simmel (1907/1971)—is that "every interaction is properly viewed as a kind of exchange" (p. 43). Moreover, exchange in some sense creates value, Simmel said:

> What one expends in interaction can only be one's one energy, the transmission of one's own substance. Conversely, exchange takes place not for the sake of an object previously possessed by another person, but rather for the sake of one's own feeling about an object, a feeling which the other previously did not possess. The meaning of exchange, moreover, is that the sum of values is greater afterwards than it was before and this implies that each party gives the other more than he had himself possessed. (p. 44)

The interaction inherent in "The expert" is essentially an exchange of information from the person asked (the expert) to the person who is asking (the nonexpert). With respect to the information exchanged, it is an unequal interaction. The person providing information normally cannot expect to get information of equal value in exchange. This is true for all the situations cited earlier. Neither the person asked for directions to the station nor the doctor can expect the person who is asking them to provide any interesting information in exchange.

The Value "Truth." A specific value is created in "The expert"-interactions: truth. If the person asking cannot presuppose that the expert's answer is true—as far as truth is possible for the involved question—the whole form of interaction with "The expert" would be senseless. Truth is a necessary presupposition of "The expert"-interaction. Asking for directions to the station would be nonsensical without presuming that there would be—although there is a station—no way of getting to the station or that it would be impossible to speak about directions to the station. Truth in this sense means the simple Aristotelian–Tarskian *adequatio rei et idearum*—the correspondence between things and ideas. A sentence such as "This or that is the shortest way to the station" is true if (and only if) this or that is the shortest way to the station. Truth is an option that can be realized in every interaction with experts.

When we say that the interactional form "The expert" creates the value "truth," this does not mean that experts, in general, tell the truth. Nor does it mean that every client or person asking expects the expert to tell the truth. It also does not mean that truth exists in a metaphysical sense. We also do not stake the claim that the value of truth is exclusive to "The expert"-interaction; there might be other occasions or institutions that create truth, such as scientific or legal discourses. As already said, truth is a presupposition of "The expert"-interaction. Even if someone who is generally suspicious of scientific knowledge—be it modern medicine or political sciences—asks a scientist about the current trends or information on the state of the art, this person would nevertheless expect the scientist to truthfully explain the state of the art in that particular science.

How can interaction create the value "truth?" According to Simmel (1907/1971), the interactional forms may be seen as settled conflicts or "peace treaties" (p. 67). Interaction is exchange, and in general there is no true value for the objects that are exchanged; the values are determined by the desires of the person. The exchange value has to be negotiated. As to information and explanation, truth is a stabilized settling point. An explanation cannot be more than "true." Truth and validity of information might be put into question in scientific discussions, but normally not within "The expert"-interaction.

Evidence From Applied Linguistics. Support for understanding "expert" as a social form comes from modern applied linguistics. From the point of view of applied linguistics, the distribution of expertise in interaction has to be seen as a jointly constructed achievement between the participants (see Ochs, 1991; Schegloff, 1991). An empirical study on the "constitution of expert-novice in scientific discourse" that used conversational analysis demonstrated some basic features of the expert's role (Jacoby & Gonzales, 1991):

- the dual and relative character of being expert in relation to a non-expert: "The constitution of a participant as expert at any moment in ongoing interaction can also be a simultaneous constitution of some other participant (or participants) as less expert, and [. . .] these interactionally achieved identities are only candidate constitutions of Self and Other until some next interactional move either ratifies or rejects them in some way" (p. 149);
- the phenomenon of shifting expertise: "The same individual can be constituted as an expert in one knowledge domain, but constituted as a novice when traversing to some other knowledge domain. Secondly, within

a single knowledge domain, the same individual can be constituted now as more knowing, now as less knowing. Finally, in either of these two situations, the valence of expertise may shift with a change of recipients." (p. 168)

To summarize the essentials of the concept: "Expert" refers to an interaction rather than a person; it is a social form. "The expert"-interaction can be defined by somebody explaining a matter (what, how, and/or why) to someone else. Inherent in this form of interaction is the value of truth: From a theoretical point of view, this means that truth is a presupposition of "The expert"-interaction. From a pragmatic point of view, discussion with an expert (within "The expert"-interaction) comes to an end when the expert states what he or she knows as the "true" facts.

The following two paragraphs view the concept of "The expert" as a social form from two different points of view. To link the concept to basic sociological theory, we first introduce Mead's concepts of role-taking. The second paragraph examines the experts' role in courts and reviews the concept of "expert" as a social form for judicial systems.

Experts From the Point of View of Role-Taking

Of course, "expert" in some sense denotes a role. We could also say that experts have a role—but then we would presuppose a dichotomy between expert individuals and the roles they can assume. This could end in reifying "experts" and "roles," leaving us without an understanding of the flexibility of the use of the form "The expert." Instead, we should try to understand the sociopsychological basis of role-taking that is present when someone is addressed or acts as expert. The basic concept of role-taking was introduced by the American sociologist and philosopher George Herbert Mead.

The Basic Concepts "I," "Me," and "the Generalized Other." Mead's concept of role-taking can be seen as an explication of his fundamental contention: "We must be others if we are to be ourselves" (Mead, 1924/1925, p. 276). Mead's concept is based on an analysis of the structure of the human self. Mead distinguished between two components, the "I" and the "me"; the "me" is the controlling, societal component of a person's behavior, whereas the "I" is the impulsive component of behavior, on which the "me" judges. In Mead's (1977) words, "The 'I' is the response of the organism to the attitudes of the others; the 'me' is the organized set of attitudes of others which one himself assumes. The attitudes of the others constitute the organized 'me,' and then one reacts towards that as an 'I' " (p. 230). The "I" and "me" result from the necessity of coordinating social behavior. Social behavior is coordinated by individuals learning to react in accordance to others. The coordination creates what Mead called a *social act*—an act that is distributed among several actors. "In the complex life of the group, the acts of the individuals are completed only through the acts of other individuals" (1924/1925, p. 262). Mead exemplified this idea of distributed, coordinated action through relationships in games and exchanges. Buyers and sellers do not exist other than through two-sided coordination. The same is true for games such as baseball or chess. The coordination requires that parts of the whole social act are present in the form or structure of each individual's behavior. Mead (1977) conceived the central form of social coordination as the *generalized other*: "The organized community or social group which gives to the individual his unity of self can be called 'the generalized other' " (p. 218).

The "generalized other" exists through coordinating mutual individual role-taking. For the individual, social coordination creates the two components of the self: on the one hand the "me" that is the product of coordination, on the other hand the "I" that has to be coordinated.

> It is just because the individual finds himself taking into account the attitudes of the others who are involved in his conduct that he becomes an object for himself. It is only by taking on the roles of others that we have been able to come back to ourselves [. . .] the self can exist for the individual only if he assumes the roles of the others. (1977, p. 268)

Social Coordination and Social Control With the Expert Role. We now turn to the question of what it is to be an expert or to consult someone as expert from the point of view of role-taking. Again we have to distinguish the form of the role or interaction called "The expert" and the content—that is, the various persons providing specific expert knowledge or services. As to the aspect of social form (or the role), Mead (1977) spoke of "abstract social classes or subgroups, such as the class of debtors or the class of creditors" (p. 221). These abstract aggregations define possible forms of coordination among individuals. They "afford or represent unlimited possibilities for the widening and ramifying and enriching of the social relations among all the individual members of a given society as an organized and unified whole" (loc. cit.). The social form "The expert" simply defines a class of possible relations among members of a society, thus being part of "the generalized other."

The concrete expert has to be seen in the context of the coordinated social act in which "The expert"-interaction takes part (or can take part). From the point of view of a particular person addressed as "expert," the expert role defines the person's "me," whereas the "I" of this person is determined by all his or her personal circumstances of this "The expert"-interaction—the person's knowledge, experiences, motivations, and so on. In so far as experts are sought because of their knowledge, the expert knowledge is the object of the social act that involves experts. According to Mead (1924/1925), social objects exert some sort of social control because the actions of individuals have to be coordinated in accordance with the social object (pp. 274–275). If we try to improve the way in which we play chess and thus consult a grandmaster who shows us how to play the opening, this piece of "expert" information will probably influence our next game (we might even play worse). The social control through information is two-sided: Not only may the nonexpert try to put into practice the expert's advice, but also the expert has to check what kind of information can serve as expert advice. Again, roles are not impersonal forces; they are based on the necessity of coordinating social action. Roles and social control do exist insofar as coordination through role-taking takes place.

Courts, Experts, and Truth

The interaction of "The expert" implies truth. This is the reason that experts are not only of special interest for litigation and judicial processes. Furthermore, it also is responsible for special judicial precautions for expert testimony. Thus, we examine the role of experts in law.

One might object on the grounds that courts are only one specific field for experts and that law is only one subsystem of society. Thus, we might ask: Is it possible to generalize from what we know about experts in courts to a general form of "expert"? Experts are explicitly part of the judicial system; thus, the social form "The expert" should be present in judicial processes, too. Moreover, modern judicial systems have become very formalized. Irregularities in judicial processes are subject to revision. Thus, when looking for any definitions for experts besides what academic scholars think about experts, we are well advised to study definitions for the expert's role in legal systems.

We consider both forms of modern juridical systems: the common law approach of Anglo-American courts as well as the statutory law approach dominating in continental Europe. The use and significance of experts differs between these two judicial approaches. We first consider experts in U.S. courts because of the traditional significance of juries to U.S. courts. Second, we look at the use of experts according to the German rules of procedure to show an example of continental juridical systems.

Common law is "judge-made, bench-made law rather than a fixed body of definite rules such a the modern civil law codes" (Abraham, 1993, pp. 6–7). "Often based on precedents, common law embodies continuity in that it binds the present with the past" (p. 7). The common law tradition originated in England; characteristically, England has no Constitution. In contrast to common law, statutory law is a written and enacted system of laws; statutory law

> originates with specially designated, authoritative lawmaking bodies—presumably legislatures, but also executive-administrative decrees and ordinances, treaties, and protocols, all of which are committed to paper. (p. 14)

The two outstanding models for modern law systems in continental Europe were the *Corpus Juris Civilis* of the Roman Emperor Justinian I (527–565) and the Code Napoleon codified by the Emperor Napoleon I at the beginning of the 19th century.

In U.S. courts, experts are quite common and play an important role. The use of experts in trials is regulated by Federal Rules of Evidence 702 through 706. Rule 702 introducing the expert reads:

> If scientific, technical, or other specialized knowledge will assist the trier of fact to understand the evidence or to determine a fact in issue, a witness qualified as an expert by knowledge, skill, experience, training, or education, may testify in the form of an opinion or otherwise.

There is no restriction as to academic or professional qualifications the expert should have. In the case *United States v. Johnson* (1978/1979), an expert was allowed to identify the marijuana involved as having come from Colombia, his expertise originating in having smoked marijuana over a thousand times (see Rossi, 1991, p. 7). Nevertheless, as we discuss, professional qualifications play an important role in trials.

A U.S. trial is an adversary process. As Robert H. Jackson, U.S. Supreme Court of Justice, stated: "A *common law* trial is and always should be an adversary proceeding" (Abraham, 1993, p. 96). There are always two parties involved, each of them trying to get the most favorable result by *knocking out* the adversary party. Many U.S. lawyers would have difficulties imagining a trial as anything else but a competition, sometimes viewed as a grim game (Abraham, 1993). This also determines the role and use of experts within U.S. trials. It can be characterized through three aspects.

First, experts are witnesses. As a witness, an expert is obliged to testify knowledge he or she has obtained through personal experience (Federal Rule of Evidence 603; see Mueller & Kirkpatrick, 1993). In general, hearsay evidence is not admissible (Rule 802). The obvious problem arising is that most scientific knowledge experts may base their opinions on what is normally taught through academic hearsay. Therefore, Rule 703 concedes:

> The facts or data in the particular case upon which an expert bases an opinion or inference may be those perceived or made known to the expert at or before the hearing. If of a type reasonable relied upon by experts in the particular field in forming opinions or inferences upon the subject, the facts or data need not be admissible.

Whatever the empirical basis may be, the expert as a witness is expected to testify facts. In practice, the lawyer has to instruct the expert:

> The lawyer's concept of fact is very different from the layman's. When the expert who has not been sufficiently briefed in his role is asked to give his opinion, he usually thinks that you are asking him for truth. If he is honest, he will begin to worry about whether he is cheating intellectually when you press him to assert his inferences as facts. (Watson, 1973, p. 79)

Second, experts—commonly—are partial. Each party tries to find an expert who can testify in its favor. Consequently, contradicting experts are common in U.S. courts. "In two thirds of the trials with expert testimony (57% of all trials), there were opposing experts in the same general area of expertise—most often opposing medical experts" (Allen & Miller, 1993, p. 1140).

In a study on expert witnesses, a judge complained: "the 'swearing contests' that take place between expert witnesses are a national disgrace" (Cecil & Willging, 1993, p. 13). Even if the parties' experts only promote their personal opinion, they are partial through selection. Therefore, prohibiting testimony of experts of "junk science" would not have any effect—not to mention the judicial and epistemological problems of defining *junk science* (Huber, 1993). From the point of view of a lawyer, it is advisable to prefer experts who believe what they testify:

> No matter who appoints the expert or what the function of his testimony, he will always be a partisan for his own opinion. He certainly should believe his opinion and be able to argue it vigorously. While that opinion may not be the ultimate fact or truth, it should reflect what he believes to be true. (Watson, 1973, p. 79)

Third, experts are subject to cross-examination in the presence of a jury. In the jury system, justice is realized by the verdict of an impartial jury. "There is no doubt that the institution of the jury is at once one of the most fascinating and one of the most controversial aspects of the judicial process and that justice in the West owes a great debt to the jury system" (Abraham, 1993, p. 100).

The jury system is based on the fundamental belief that every piece of reasonable evidence in a case can also be evaluated by a jury of average reasonable persons who are provided with the necessary information to do so. Therefore, the following aspects of using experts are left to cross-examination by the lawyers:

- What qualifies a person to become an expert?
- What data forms the basis for an expert's opinion?
- How does an expert depend on a party?

It is in the interest of each party to examine the experts of the adversary party and demonstrate lack of evidence if this proves to be the case. As one judge put it, "[T]he lawyers are pretty good about shooting holes in each others experts. It's generally a credibility question and the jury can sort it out" (Cecil & Willging, 1993, p. 21). In accordance to Rule 702, the main admission criterion for an expert is helpfulness. The expert can testify on almost anything that is helpful and relevant to the trial (Rossi, 1991, chap. 2).

Experts in U.S. trials, as introduced here, are a helpful means in the hands of the parties involved. The adversary process prevents the jury from undue deference to experts. To some extent, experts—as witnesses in general—are used up in the adversary process, in particular in cross-examinations. A lawyer resumes: "Another major area [besides compensation] I talk over with my expert is what he fears most—cross-examination [. . .] I have never met an expert of any kind who has any liking for lawyers" (Shepherd, 1973, p. 21).

However, there is an alternative to the experts hired by the parties to a trial: the court-appointed expert. Federal Rule of Evidence 706 allows the appointment of experts: "Underlying this authority is the broader inherent authority of the court to appoint experts who are necessary to enable the court to carry out its duties. [. . .]" A survey by the Federal Judicial Center (1993) asked, "Why are court-appointed experts, as authorized by Federal Rule of Evidence 706, employed so infrequently?" (Cecil & Willging, 1993, p. 3). According to this survey, 87% of court judges considered court-appointed experts helpful (p. 19, Table 2), and only 20% of judges ever had appointed an expert, most (52%) having done so only once (p. 8).

Appointment of experts through the court transcends the adversary system of common law. Thus, the general form of the interaction "The expert" comes into play again, as well as the consequences of consulting an expert. Appointed experts are considered to be "reserved for cases with extraordinary needs" (Cecil & Willging, 1993, p. 18). The most prominent reason for failure to appoint an expert is the "infrequency of cases requiring extraordinary assistance," as the survey found. Another important reason is "respect for the adversary system" (p. 20): "We're conditioned to respect the adversary process. If a lawyer fails to explain the basis for a case, that's his problem" (p. 21).

The problem with court-appointed experts is the mitigation of the unwelcome consequences of consulting an expert, especially the value of truth attached to "The expert"-interaction that is normally eliminated in the adversary process. Thus, the Advisory Committee notes accompanying Rule 706 warn that "court-appointed experts acquire an aura of infallibility to which they are not entitled" (p. 55), one court concluding that a court-appointed expert "would most certainly create a strong, if not overwhelming, impression of 'impartiality' and 'objectivity' [which] could potentially transform a trial by jury into a trial by witness" (p. 48). The survey, too, revealed that "juries and judges alike tend to decide cases consistent with the advice and testimony of court-appointed experts" (p. 52).

> The most dramatic illustration of dominance by a court expert occurred in a case in which a large number of workers claimed damages due to working conditions. At the behest of the court, a physician examined all of the workers and reported findings for each plaintiff. The physician's court-appointed status was disclosed to the jury, and the judge reported "the jury discounted the experts for each side." In fact, in each individual case, the jury followed the findings of the court-appointed expert, finding sometimes for the plaintiff and sometimes for the defendant. (p. 54)

Thus, in case of court-appointed experts, the relationship between the roles of the judge and the expert needs clarification, "distinguishing between the expert's duty to provide technical expertise and the judge's duty to decide the case" (p. 36). For example, one judge said concerning a court-appointed computer expert: "I instruct [the expert] that his role was to help me and that he was not to decide the case. His main role was to interpret the language to me, give me the background on computer technology, tell me how the various systems work."

Statutory judicial frameworks, such as in France and Germany, tend to systematize and differentiate the legal matters, including actors and functions. In statutory systems, the judge normally plays a more active role, often taking over the examination of witnesses. Within the borders of the legal system, the judges are given full discretion. Juries are rarely used; however, we sometimes find them in criminal courts. In statutory law courts, the judges are the main acting characters, not the adversary parties.

In statutory systems, the role of the expert in the court is more systematized than in common law. In German law, for example, there is a clear distinction between witnesses and experts. Witnesses have to have personal evidence of the persons and events that caused a particular case (e.g., in the case of a traffic accident). Witnesses are literally eye witnesses or persons with some relevant information gained at the time of, or shortly before, the event. In contrast, an expert (Sachverständiger) functions as "personal means of truth" (persönliches Beweismittel). The expert is considered as a "judicial clerk" (Gehilfe des Gerichts) or "clerk of the judge" (Richtergehilfe) who, under the supervision of the judge, helps the court interpret and understand a case. More precisely, the task of the expert according to a standard commentary on experts is threefold:

1. Inform the court about principles derived from experience in his/her area of expertise.
2. Determine facts on the basis of his/her expertise.
3. Judge facts on the basis of the principles derived from experience in his/her area of expertise. (Jessnitzer & Frieling, 1992, trans.).

As is the case with court-appointed experts according to U.S. Federal Rule of Evidence 706, there is a clear division of roles in German law: The expert informs and the judge decides. We call this the *information clause*.

When introducing the expert, the German Rules of Procedure are much more unclear than the Federal Rules of Evidence. Nothing is said about the knowledge or qualifications necessary to qualify as an expert. The appointment of an expert in a civil case is regulated by:

> § 404 Appointment. [i] The appointment of experts and the number of experts is in the responsibility of the court hearing the case. (trans.)

The Rules of Procedure for experts in criminal cases (§§StPO 72-92) are as vague as the ones for experts in civil cases (§§ZPO 402-414). The French *Nouveau code des procédure civile* simply reads, "La decision qui ordonne l'expertise [. . .] nomme l'expert ou les expert" (in short: the expert or the experts are to be appointed). Given the German Rules of Procedure, a judge could appoint any kind of expert. Mostly, the judge uses publicly appointed (*öffentlich bestellte*) experts—a qualification one can apply for—or members of professions, avoiding reasons for appellation as to the expert's qualification. In fact, depending on the case, there can be a variety of experts appointed—mostly physicians and experts on construction, insurance, and automobiles, but also astrologers, occultists, and pendulum specialists (Dippel, 1986).

The appointed expert normally has to prepare a report (*Gutachten*) for the court. The report is based on case-specific questions prepared by the judge. The judge has to decide whether the report is convincing (Baumbach, Lauterbach, Albers, & Hartmann, 1993). The expert normally presents the report during the proceedings and can be sworn in. There are no specific regulations for the report, but the judicial function of the report implies several demands (Wellmann, 1988):

- the content should be comprehensible, also to a third party not familiar with the case;
- the inferences should be conclusive to every reasonable person.

In other words: Any reasonable person investing the same amount of time studying a specific subject should come to the same conclusions as the expert. In the remainder of this book, we call this the *generalizability clause*.

Both judicial forms of processes—the adversary process with a jury and the process according to the differentiated continental law—take precautions against undue influences of experts. On the one hand, we have the confrontation and cross-examination of experts, the expert being a witness with special knowledge subject to examination just like other witnesses. On the other hand, we have an expert preparing a report under the supervision of a judge, the expert having a well-defined function in the judicial proceedings. In contrast to witnesses, according to German law, experts in principle have to be *vertretbar*, which means of a kind that they can be substituted for one another (Baumbach et al., 1993, p. 1178). From this point of view, an expert is obliged to only report standard knowledge. A report using new diagnostic criteria or based on a revolutionary theory of diagnosis would not meet the required standardization. We call this principle the *standardization clause*.

The expert, as defined by the differentiated continental law, can be considered as a viable extended formalization of the interaction of "The expert." From this point of view, an expert

- reports information relevant to a certain question involved in a case, but refrains from deciding cn the case (information clause),
- reports in a manner that makes it possible for any reasonable person investing the same amount of time to obtain the necessary experience and knowledge to arrive at the same conclusions (generalizability clause), and
- reports standards of specialized knowledge (standardization clause).

In summary, "The expert" refers to an interaction rather than a person; it is a social form. Inherent in this form of interaction is the value of truth; truth is a presupposition of "The expert"-interaction. Thus, in an interaction with an expert, the examination of a certain case comes to an end when the expert says what the true facts are. Courts in both types of legal framework—the adversary process with a jury and the process according to the differentiated continental law—take precautions against undue influences through experts. Whereas experts are special witnesses in the adversary type, differentiated continental law defines a special role for experts, thus providing an extended formalization of the interaction "The expert."

3.2 ATTRIBUTION: "THE EXPERT" IS A SOCIAL FORM THROUGH WHICH SOMEONE IS ATTRIBUTED AS "EXPERT"

Usually we do not think of an expert as an interaction. Usually experts are considered as persons who are—in a certain aspect—qualified. However, to assume the role of an expert, a qualified person has to be recognized as an "expert"—at least by one other person. In the preceding paragraph, we introduced the concept of "expert" as a social form of interaction. Now we turn to this question: How and why do we identify experts in such interactions? To this end, we introduce attributional theory. We see that regarding somebody as an expert is an attribution that follows the same rules as common social cognition of other persons.

Personal Causality—A Short Introduction into Attributional Theory

In 1958, the psychologist Fritz Heider published a book on *The Psychology of Interpersonal Relations* that founded a new research program known today as *attributional theory*. The starting point for Heider was this question: How does a person interpret the actions of another person? Thus, he started by investigating commonsense psychology. Heider wrote:

> The study of common-sense psychology is of value for the scientific understanding of personal relations in two ways. First, since common-sense psychology guides our behavior toward other people, it is an essential part of the phenomena in which we are interested. In everyday life we form ideas about other people and about social situations. We interpret the actions of other people and we predict what they will do under certain circumstances. Though these ideas are usually not formulated, they often function adequately. (p. 5)

Heider compared social perception to perception in general. One phenomenon Heider found in perception in general as well as in social perception was the importance of invariants: invariant characteristics such as the right angles of a chair or, in people, invariant traits of personality. Heider spoke of *dispositional characteristics*:

> It is interesting that in social perception, also, the direct impressions we form of another person, even if they are not correct, refer to dispositional characteristics. At least, relative to the events that mediate these impressions, the characteristics show a high degree of intrinsic invariance. For instance, the impression that a person is friendly, which may be conveyed in any number of ways, points to a relatively enduring characteristic of the person. (p. 30)

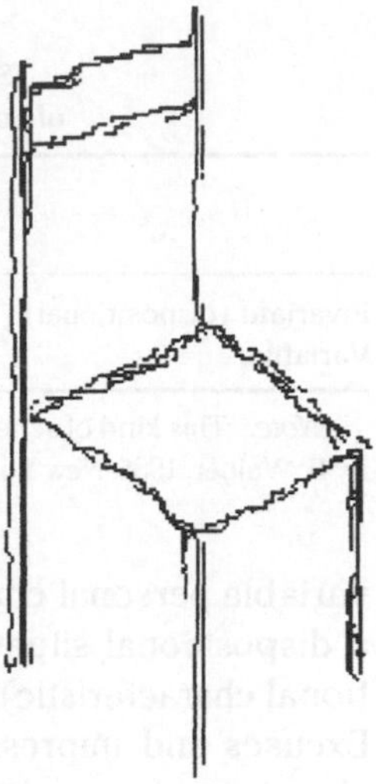

FIG. 3.1. Presupposing invariant characteristics: drawing of chair. There is not any right angle in the drawing, but what we see is a chair that has right angles between its constructional elements.

Another phenomenon that social perception and perception in general have in common is the presupposition of causal connections. The French psychologist Albert Michotte demonstrated that there is a strong tendency to interpret or perceive in terms of causality (1946/1963). For his experiments, he used revolving discs that produced the impression of moving points. Depending on speed, direction, and form of the moving points, they are seen as influencing one another: launching, blocking, pursuing, and so on, one point being a moving "actor," the other one being a moved "victim."

In social perception, too, a person's behavior is interpreted in causal connections. Especially behavioral consequences are attributed to supposed causes. Attributional theory basically distinguishes two main, dichotomous sources of attributed causality: the person or the situation. Both of them have invariant (dispositional) or variable properties. For instance, Kathy, age 9, passed an exam with great success. As Table 3.1 shows, there are at least four possible causes: She is talented (factor ability, a dispositional personal characteristic), she has learned a lot (factor effort; a

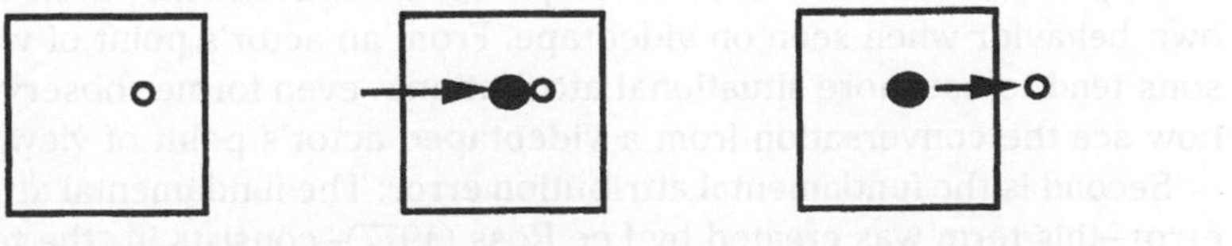

FIG. 3.2. Presupposing causality. Three scenes from an experimental procedure: 1. a white spot in a field; 2. a black spot arrives and approaches the white one; 3. the white spot is launched. In the sequence of this scenes, the black spot seems to have caused the movement of the white spot.

TABLE 3.1
Sources for Causal Attributions in the Case of Success or Failure (e.g., in an achievement test)

	Person (Internal)	*Situation (External)*
Invariant (dispositional)	Ability	Task difficulty
Variable	Effort	Luck

Note. This kind of scheme was introduced in *An Attributional Theory of Motivation and Control* by B. Weiner, 1986, New York: Springer. Copyright 1986 by Springer. Reprinted by permission.

variable personal characteristic), the exam was easy (factor task difficulty, a dispositional situational characteristic), or it was luck (a variable situational characteristic). In this way, responsibility and prestige are attributed. Excuses and impression management have to operate with attribution—claiming a success as proof of one's giftedness (factor ability) and blaming others in case of failure (situational factors). Psychological research on conditions and consequences of human attribution abounds. We describe two characteristics of the common attributional process in more detail. First is observer–actor discrepancy: Actors and observers do not always arrive at the same explanation of the actor's behavior. Jones and Nisbett (1971) argued that actors who explain their own behavior are inclined to give considerable weight to situational causes. Observers place more emphasis on dispositional causes of the actor's behavior. Michael D. Storms (1973) demonstrated that this can be true for changed positions: A person observing his or her own videotaped actions takes a more dispositional explanation of his or her own behavior.

In Storms' experimental setting, two persons have a getting-acquainted conversation, and a third person is the observer (see Table 3.2). Two types of videotapes are recorded—one showing the conversation from the observer's point and one showing the point of view of one of the actors in the conversation. The actor has to explain his or her own behavior, and the observer has to explain the actor's behavior. From the observer's point of view, persons tend to use more dispositional attributions—even for their own behavior when seen on videotape. From an actor's point of view, persons tend to use more situational attributions—even former observers who now see the conversation from a videotaped actor's point of view.

Second is the fundamental attribution error: The fundamental attribution error—this term was created by Lee Ross (1977)—consists in "the tendency for attributers to underestimate the impact of situational factors and to overestimate the role of dispositional factors" (p. 183). Heider (1958), too, mentioned this error:

TABLE 3.2
Emphasis on Dispositional (Personal) Causal Attribution Depending on the Role as Actor or Observer

		Experimental Conditions		*Experimental Results*	
	What Is Asked?	*Original Position*	*Changed Position*	*Original Position*	*Changed Position*
Actor	Attribution of own behavior	Acting in the conversation	Sees own behavior (videotaped)	0.2 ("Situational")	6.8 ("Dispositional")
Observer	Attribution of actor's behavior	Sees the actor	Sees situation (videotaped from actor's point of view)	4.9 ("Dispositional")	1.6 ("Situational")

Note. The table shows results from "Videotape and the Attribution Process: Reviewing Actors' and Observers' Points of View" by M. D. Storms, 1973, *Journal of Personality and Social Psychology, 27,* Table 1. Copyright 1973 by American Psychological Association. Reprinted by permission. Actors and observers had to change their positions in a getting-acquainted conversation. The values mean the average use of dispositional causal explanations (0 through 9), high values indicating the dominance of dispositional causal explanations, low values indicating the dominance of situational explanations. Attributions of actors who saw their own behavior on videotape were more dispositional (6.8) than those of actors in the conversation (0.2).

> [T]he father always sees his son in the role of son, the employer sees the employee only as an employee behaving in front of the employer etc [. . .] it is as if we always carried a flashlight with a filter of red color when examining an empty room; we would then ascribe the color to the room. (p. 55)

This error of overestimating dispositional factors (such as personality traits) and neglecting situational factors (such as role relationships or group influences) is quite common in everyday commonsense psychology. Ross (1977) said the commonsense psychologist "too readily infers broad dispositions and expects consistency in behavior or outcomes across widely disparate situations and contexts" (p. 184). In fact, dispositions such as friendliness, honesty, ambition, or miserliness cannot explain more than 10% of the specific behavior in different situations, as Mischel (1968) showed in a compilation of psychological research on personality traits.

The fundamental attribution error also appeared in Milgram's (1963) electroshock experiment. In his experiments, 65% of the invited participants—all of them *normal* people—on command went as far as to induce deadly 450-volt electroshocks on other participants. Neither the participants nor the psychological experts could expect the amount of obedience to authority. Expert psychiatrists predicted that only about 1% of the subjects in the Milgram experiment would participate until the end—the 450-volt condition (Milgram, 1974). Even in a faithful verbatim reenactment of the Milgram experiment, participants, regardless of whether they played the role of a subject in the reenactment or merely observed, totally underestimated the situational impacts and compliance of Milgram's participants (Bierbrauer, 1979).

The Need for Controllability and Accountability

Attribution theory explains the common tendency for personal attributions. Effects and events—acceptable ones such as genius inventions or political reforms, or unacceptable ones such as crimes and traffic accidents—are attributed to persons mostly through dispositional causal attributions. Therefore, persons are "inventors," "reformers," "criminals," "bad risks," or "experts." It cannot be a surprise that information and explanations provided by an expert are easily attributed to a stable dispositional property of that person—his or her expertise.

Causal attributions to invariant dispositional properties "make possible a more or less stable, predictable, and controllable world" (Heider, 1958, p. 80). Even if attributions of personal invariants—such as many personality traits—might be incorrect, they function in certain situations. Usually we are not engaged in social situations that change from day to day: We meet many people only in specific situations—some as friends, others as colleagues, shop keepers, newspaper sellers, and so on. It is a reasonably good guess to assume that they will not change their behavior. For many reasons, it might be a reasonably good guess even in new situations.

In a series of experiments, Langer (1983) demonstrated the impact of the need for controllability. The phenomenon she discovered was the *illusion of control*. Even in a strictly randomized situation, such as participating in a lottery, people might imagine that they have control over the outcome. In an experiment, $1 lottery tickets were sold to office workers. Some were simply handed the ticket, whereas others had to choose a ticket. Several days later, the subjects were approached and asked to resell the tickets. The office workers who had had the possibility to choose their tickets sold them at an average price of about $9; the others who had not had this option sold their tickets for about $2 (Langer, 1983). Langer showed the illusion of control for many other conditions besides choice; under lotterylike conditions, subjects have the illusion of control when:

- they compete with a downgradable competitor,
- they are to a high degree actively or passively involved in the situation, or
- the situation seems familiar.

It goes without saying that, also in sciences, as H. M. Collins argued, there is the strong tendency to find regularities that can be interpreted as causal relations. The scientist who fails to find such regularities risks being considered as not competent: "We perceive regularity and order because any perception of irregularity in an institutionalized rule is translated by ourselves and others as a fault in the perceiver or in some other part of the train of perception" (Collins, 1985, p. 147).

The need for a predictable and controllable world and the prevalence of personal causal attributions result in the fact that important parts of our social life—from car accidents to political reforms—can be accounted to persons. Accountability serves social regulation. Thus, a car accident is regulated by finding the one who is to blame. When facing problems, it is easier to dismiss an employer or discharge a minister than to change working conditions or general policies. Thus, we can say that social life—to some extent—is regulated by personal attribution. As the social psychologist Tetlock (1992) put it: "The accountability of conduct is a sociocultural adaptation to the problem of how to coordinate relationships among individuals who are capable of observing, commenting on, and controlling their own actions" (p. 337).

Experts are accountable for reliable information and explanation. This is true for the everyday expert on the street who is asked for directions to the station as well as the physician or person speaking as a court-appointed expert on asbestos. The accountability is constituted by the interaction of the

"The expert"-form. In the outset of this chapter, we referred to Simmel, who considered every interaction as interpersonal exchange and reciprocal influence. We could describe "The expert"-interaction in terms of negotiating accountability:

(a) the nonexpert makes the expert accountable for the validity of the information and explanation offered by the expert;
(b) the expert copes with being accountable, one effect of coping being that the accountable person sticks to "acceptable" reactions (Tetlock, 1992).

When we regard "expert" as a form of interaction, we see that it is indeed possible for the person addressed as an "expert" to reject being "The expert." We discuss this phenomenon in the context of the expert's autonomy.

> We could also describe "The expert"-form of interaction in terms of *trust*. Usually—but not necessarily—what the nonexpert invests in the "The expert"-interaction is trust. The nonexpert asks someone whom she believes to be qualified to know the answer. This is a type of trust in a person that is involved in many kinds of interactions and that may be seen as a product of personal causal attributions.

Also Simmel (1907/1971) said about trust: ". . . not something, but someone is believed" (p. 66).

Now we can say more about the attribution processes in "The expert"-interaction. For example, what is to say that the information and explanation provided by an expert is attributed to a stable dispositional property of that person—namely, his or her expertise? In a first approximation, we can distinguish several sources of expertise. Usually we regard expertise as based on experience, expertise being a personal dispositional characteristic. As Table 3.3 shows, however, we can also infer expertise from the fact that a person is a member of a profession. This is the case when our sole information about a person concerns his or her profession: He or she is a lawyer, or physician, or an academic. There might also be a case where a person addressed as an expert has the requested information by chance. This would be the case when we ask someone for directions to the station and the person whom we ask is also a stranger, but the sole information about the town he has concerns the directions to the station. Then we attribute expertise to the person, but as a variable capability.

Table 3.3 does not explain differences in attributed accountability. Even if expertise is commonly regarded as based on experience and not on membership of a profession, professionalism seems to imply more accountability than expertise through simple experience. Figure 3.3 shows how the as-

TABLE 3.3
Sources for Attributing Expertise to a Person

	Internal (Person)	External (Situation)
Dispositional	Experience, knowledge	Profession
Variable	Having information	Vote, guess, etc.

Note. See also Table 3.1. The common attribution is experience as a dispositional personal attribute. The cells only show examples for attributions; for instance, membership to a profession is not the only external attribute that might inform us about expertise, another example being the uniform of a policeman. To call someone an "expert" in most cases is a personal dispositional attribution; external or variable attributions are less common. There might also be cases where a person is expert by vote or that expertise is based on guessing "facts"—but these cases do not seem quite realistic—even a clairvoyant doesn't guess; clairvoyance is seen to be based on dispositional "supernatural" capabilities.

pects of attribution and types of expertise can be linked. We have three important aspects of attribution:

(1) Attribution is personal: That is, we normally address a single person as an "expert" (and not a constellation between persons).
(2) Attribution is dispositional when we attribute the expertise of the expert to his or her stable capabilities.
(3) Attribution is causal when we see experts and effects as causally linked; for example, a doctor's treatment causes the patient to recover.

As Fig. 3.3 shows, "expert" is a simple personal attribution. Adding a dispositional aspect, we can speak of experienced persons. For professionals

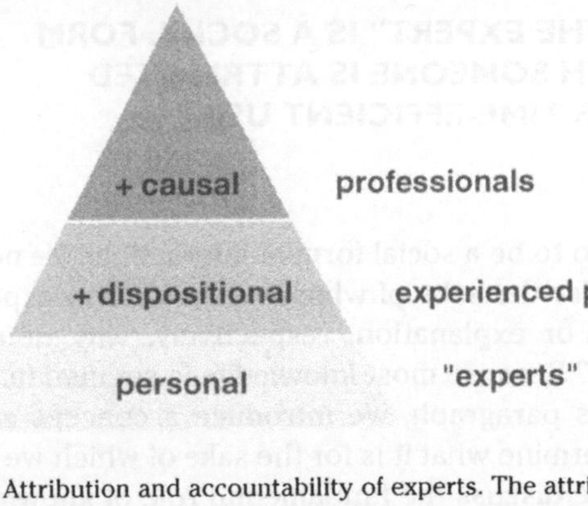

FIG. 3.3. Attribution and accountability of experts. The attribution of professional has three aspects: causal, dispositional, and personal. To speak of an "expert," only personal attribution is necessary; in this case, accountability of the "expert" is reduced, too.

who also provide treatment, attribution has to be causal. We can say that each attributional aspect adds to accountability, "experts" through simple personal attribution being less accountable than professionals. Thus, when we speak of experts, we have to distinguish among

(1) the social form "The expert," a form of interaction that provides information and explanation and makes the presupposition of truth necessary;
(2) the person who is actually addressed (attributed) as an "expert" in interaction (1) This can be "any" person on the street (who knows and shows me the directions to the station);
(3) the person who is qualified through his or her expertise to be addressed as an expert in sense 2;
(4) the person who is usually or professionally engaged as an expert in sense 2. These are, for instance, physicians, lawyers, and scientists or academics, respectively.

In the remainder of the book, we speak of experts in sense 2.

To summarize, commonsense psychology prefers personal dispositional attributions. Thus, it attributes the outcomes of "The expert"-interaction–mainly information–to a stable personal source–the expert's knowledge or experience, respectively. Personal attributions preserve the accountability of social phenomena–good or bad–to persons. "The Expert"-interaction serves the need for a controllable world. The expert can be obliged to (the value of) truth.

3.3 *LEISTUNG:* "THE EXPERT" IS A SOCIAL FORM THROUGH WHICH SOMEONE IS ATTRIBUTED AS "EXPERT" FOR TIME-EFFICIENT USE OF KNOWLEDGE

"The expert" was said to be a social form of interaction. We now turn to the question: What is it for the sake of which we consult an expert? If experts provide information or explanation, respectively, why don't we consult knowledge "as such"? Because most knowledge is codified in books, we are free to use it. In this paragraph, we introduce a concept called *Leistung*, which shall help determine what it is for the sake of which we consult an expert. In particular, we discuss the fundamental role of knowledge in expert services.

From Functions to *Leistung*

In sociology and psychology, there are several prominent types of concepts for expert service. Although there are many, we discuss only three types of concepts, each of them presented in a pure, extreme version.

Function. In the works of Talcott Parsons (1939, 1951, 1968), we find the concept of function. Parsons believed that professions had to serve a responsible social function. Professions such as medicine or law are occupational groups that render competent services. According to Parsons (1968), one important criterion for the definition of a profession is "a full-fledged profession must have some institutional means of making sure that such competence will be put to socially responsible uses" (p. 536). Medicine seems to serve the health of the people; lawyers seem to function as some kind of servants to justice.

The functional view (see also Goode, 1957, 1969) was strongly opposed by the "power approach," for instance, by Terence Johnson, who conceived professions as monopolistic power groups; he considered professionalism "a peculiar form of the institutionalized control of certain occupational activities in which an occupational community defined client needs and the manner in which these were catered for" (1977, p. 105).

In functional explanations, "the consequences of some behavior or social arrangement are essentially elements of the causes of that behavior" (Stinchcombe, 1968, p. 80). Health would be a built-in intended consequence of the medical profession. Analogously, we could say that experts in general serve the function of providing information and explanation. However, the functional approach cannot grasp the dynamics of experts' roles. More important, it provides no concept for functional alternatives: Information cannot only be provided by human beings, but also by books or federal offices. If the exclusive function of experts is to inform, why do we still need them—and not use books or computers? As for experts, the concept of function needs to be reconsidered with regard to the role of knowledge.

Status. "Expert" is sometimes viewed as a special status. This view can be approached in two ways: First, "experts" are a social group defending their status; this leads to theories similar to the power approach. Because the power approach refers more to professions than to experts in general, we do not discuss it here. Second, the "expert" status is the product of specific impression management. The extreme position would be: An "expert" is merely a certain impression. For instance, the "expert" status of physicians is the result of professional and personal impression management, including white coats, an esoteric language, an aura of infallibility, and so on.

In his book *The Presentation of Self in Everyday Life*, Goffman (1959) described the art of impression management. Some of his descriptions seem to fit "expert" physicians or lawyers:

> We often find a division into back region, where the performance of a routine is prepared, and front region, where the performance is presented. Access to these regions is controlled in order to prevent the audience from seeing backstage and to prevent outsiders from coming into a performance that is not addressed to them. (p. 238)

The point to be made is that impression management cannot explain the complete "expert" phenomenon. Experts not only can take on an aura of truth, but can also be obliged to some kind of truth in actual interaction. The point is the accountability of experts for knowledge. This specific accountability, on the one hand, makes the form "The expert" useful and, on the other hand, can cause practical problems for the role and use of experts (e.g., for experts in commissions that have to come to a decision).

Person. In psychology, "experts" are considered as exceptionally experienced individuals (e.g., masters in chess or tennis or people with exceptional skills in surgery or type-writing). As already mentioned, standard work on experts, *The Nature of Expertise* (Chi, Glaser, & Farr, 1988), commented with: "How do we identify a person as exceptional or gifted?" (Posner, 1988, p. xxix). The question relates to individuals and the criteria for their expertise. Some of the main results—as already cited—are the domain specificity of expertise and the 10-years rule. Domain specificity means that expertise in one domain (e.g., Chess) need not have the same nature as expertise in another domain (e.g., Backgammon or Go). Thus, a person who has expertise in one domain does not necessarily have expertise in related domains. The 10-years rule says that the mastery of special skills and knowledge takes about 10 years to develop.

Psychologists rarely take social functions and the attribution of status into account. However, these are social conditions that also constrain the development of expertise. Some kinds of expertise (e.g., all kinds of medical know-how) can only be obtained within occupations; others, such as playing the piano, need the supporting environment of a (middle-class) family or the state. The typing machine started a gender revolution in offices: The old honorable male secretary was wiped out by the typing machine, which was more effective and created new jobs for lower paid women.

We now turn to a different notion that seems useful when characterizing the service provided by experts. To that end, we refer to Weber's (1947) definition for an *occupation* (Beruf); this term "will be applied to the mode of specialization, specification, and combination of the functions [Leistungen,

HAM] of an individual so far as it constitutes for him the basis of a continual opportunity for income or profit" (p. 250). Talcott Parsons, in his translation, uses the word *function* that Weber could also have chosen in German, but decided not to do so. Instead, Weber spoke of *Leistung*. However, in the context of occupations, *function* works out as a viable translation for *Leistung* (pl. *Leistungen*). Problems arise when Parsons has to translate the term *Leistung* and its variations in different ways. Thus, *Nutzleistungen* are utilities in an economic sense, and *Dienstleistungen* are services. The meaning of *Leistung* disappears in these translations, as does the shared underlying concept. Actually, there is no adequate one-word English translation for *Leistung*. The concept *Leistung* combines effect and efficiency; it is the specific aspect of a unit of work for which the work can be paid. Dictionaries suggest *performance* or *achievement* for *Leistung*, but in our context the definition of power in physics (in German also defined as *Leistung*) provides a useful explanation: Power (resp. *Leistung*) is work done in a certain amount of time:

Power (resp. *Leistung*) = work / time.

Using the generalizability clause, we can say: The specific *Leistung* because of which the social form "The expert" can be used is the use of an expert's compressed experience that any reasonable person could make if he or she had enough time to do so. A citizen can—most probably—give a foreigner directions to the station because he or she has been living in the town for years and can—now consulted as an expert—cognitively reconstruct the possible ways of getting there. This is an experience that the particular nonexpert could also make if he or she had lived in town for a longer period of time. The particular citizen becomes an expert for his or her town only through interaction with the foreigner who, through this interaction, attributes the "expert" status to the citizen in question. The court-appointed expert on asbestos will—most probably—have sufficient experience with asbestos. The expert on asbestos can also be seen as some kind of condensed experience anyone could make concerning asbestos with sufficient time and effort.

In general, the expert's *Leistung* is based on a long period of perception schooling and on the ability to use categorizations. Experts are fast—at least faster than nonexperts. The specific *Leistung* the expert can be used for (compressed generalizable experience) can be the basis for a continual opportunity for income or profit and thus, for some kind of occupation such as working as a physician or an academic. The *Leistung* realized by using the social form "The expert" is part of the cognitive economics of experts. Using experts is a time-efficient use of knowledge.

The concept of *Leistung* also leads us to a measurement of expertise. Insofar as expertise is based on experience, we could measure the degree of expertise through the time gain (in years). For instance, the expertise of an experienced lawyer may have a value of 20 years. We have to take into account that, even if expertise is based on experience, the development of expertise requires more than experience alone—namely, some sort of training or schooling. This is what Ericsson et al. (1993) called *deliberate practice*. A nurse with 20 years of experience may know much about patients, but might be unable to diagnose specific syndromes in the same way as a medical specialist with less patient contact but more schooling.

In the following two paragraphs, we examine the *Leistung* aspect of expert work in more detail. First, we look at how professionals work. Then we turn to the experts' *Leistung* in dealing with knowledge.

How Professionals Work

We now turn to professional work. Professionals are often referred to as "experts." Physicians and lawyers are standard examples. However, the professional's *Leistung* is not the same in every aspect as that of the expert. Usually professionals serve more functions than explaining a matter, and often they seem to be rather reluctant to really explain the matter in case. Abbott (1988) described the following sequence of professional work, diagnosis, inference, treatment (although he did not use the same sequence in describing them):

- *Diagnosis*. "[D]iagnosis first assembles clients' relevant needs into a picture and then places this picture in the proper diagnostic category" (p. 41). Professional diagnosis is therefore of a "dual nature": "Diagnosis not only seeks the right professional category for a client, but also removes the client's extraneous qualities" (pp. 40–41).
- *Inference*. Inference is a "purely professional act" (p. 40). "It takes the information of diagnosis and indicates a range of treatments with their predicted outcomes" (p. 40). "Inference can work by exclusion or by construction. Medicine, for example, tends to work by exclusion. If a case is unclear, doctors maintain a general supportive treatment while ruling out areas by using special diagnostic procedures or watching the outcomes of 'diagnostic' treatments beyond general maintenance. Classical military tactics, on the other hand, work by construction. The tactician hypothesizes enemy responses to gambits and considers their impact on his further plans" (p. 49).
- *Treatment*. "The effects of treatment parallel those of diagnosis. Like diagnosis, treatment imposes a subjective structure on the problems with which a profession works" (p. 44). However, "Just as the diagnostic sys-

tem removes the human properties of the client to produce a diagnosed case, so also the treatment system must reintroduce those properties to make treatment effective for clients, human or corporate" (p. 46). That is, each patient or client receives individual treatment.

Although Abbott did not refer to cognitive psychology, his description of professional works has parallels to the general descriptions of human problem solving that we find in cognitive psychology. In short, problem solving also follows a sequence that we can divide into three steps (Anderson, 1985; Newell & Simon, 1972):

- The *definition of a problem space*. The problem must be represented in a cognitive space. Diagnosis, too, is the representation of a problem (a sick person) within a professional categorical space (the disease entities).
- The *selection of operators*. The problem solver has to find operators or rules. We have already stressed the importance of specific heuristics (broadly speaking, rules of thumb). From the point of view of cognitive psychology, the two inferential procedures Abbott mentioned—exclusion and construction—are heuristics, too.
- The *controlled execution*. The execution is controlled by a general cognitive program that may allow for problem-specific adjustments, just as treatment does. Another parallel consists in the use of loops, which is the repetition of special steps of the problem solving process to specify the solution or the treatment effects, respectively.

Two points need to be made. First, we can perceive the importance of the representation of a problem in a (professional) problem space. The representation may in practice turn out to be purely routine. Nevertheless, professional work in general implies the interpretation of a specific problem within a professional knowledge system. Second, professionals obviously engage in more than providing information. Especially, the information clause (information without decision) generally does not hold for professionals. A doctor actually *decides* on the client's health.

Knowledge ≠ Information

Now we turn to the role of knowledge for the expert's *Leistung*. Our starting point is again professional work. Abbott's (1991) *The Future of Professions* discussed the functional alternatives to professions and professional work, respectively. Expertise cannot only be institutionalized in individuals such as professionals, but also in commodities and organizations. Commodities such as forms, dictionaries, and textbooks, as well as specialized computer systems, also provide information and explanation. The same is true for some types of organizations (e.g., large engineering firms or hospitals). They can embody knowledge that a single average person could never assemble. All these types of institutionalized expertise—professionals, commodities, organizations—can make use of the social form "The expert." If an "expert system"—a computer—provides sufficient information for interactive explanations, it may serve as an expert, too. Indeed, the explanation function was central to the original "expert system" concept as promoted by Feigenbaum (Feigenbaum & McCorduck, 1984). We show that knowledge is not simply information and that information often needs interpretation; this is the reason that we still need human experts.

Usually "experts" are defined with regard to special knowledge. The knowledge an expert provides is, for all intents and purposes, relative. It depends, on the one hand, on the social situation of the "The Expert"-interaction. A student of law may be regarded by her friends as an "expert" on legal matters—a role she would never take on among professional lawyers. In contrast, the "expert" knowledge depends on the knowledge background of *an age* (a term used by Kant). In the 16th and 17th centuries, astrology was generally regarded as being up to date. Today astrological explanations would not be generally acceptable; however, there are still astrological services—even in the shape of appointed experts at court that evaluate "good astrological practice" (see Dippel, 1986).

In the last 50 years, the philosophy of science and knowledge has changed dramatically. In the 1930s, books such as Popper's *The Logic of Scientific Discovery* (1934) or Carnap's *The Logical Structure of the World* (1928) could base evident knowledge on scientific protocol sentences (Popper) or basic data (Carnap), whereas today the idea of primary evidence is obsolete. Instead, a fact is nothing but a fact within an interpretative framework.

The main impetus within the philosophy of science came from Kuhn's (1962) book *The Structure of Scientific Revolutions*. He broke with the idea of a linear development of scientific knowledge. *Linear development* means that science proceeds by adding results of scientific research. Instead, new scientific schools and frameworks—Kuhn spoke of paradigms—compete with one another. In our context, we do not introduce the work of Kuhn, but rather the study on syphilis by Ludwik Fleck, a Polish-Jewish physician who wrote in German. It was Kuhn who brought to general recognition the work of Ludwik Fleck. As early as 1935, Fleck wrote on *The Genesis and Development of a Scientific Fact*. Fleck described the history of syphilis and showed that there is no pure fact "syphilis" as such. The concept of syphilis consists of an amalgamation of previous, historical concepts of the disease called syphilis, and its definition depends on organized scientific thought collectives that conduct paradigmatic research and propel an interpretation.

The historical sources of syphilology [the science of syphilis, HAM] can be traced back, without a break, to the end of the fifteenth century. They contain descriptions of a more or less differentiated specific disease (in modern terms a so-called *disease entity*) which historically corresponds to our concept of syphilis, although the bounds and nomenclature have undergone considerable modifications (p. 1). [T]wo points of view developed side by side, together, often at odds with each other: 1) an ethical-mystical disease entity of "carnal scourge," and 2) an empirical-therapeutic disease entity, (p. 5) the latter being related to the treatment with mercury. Further ideas were based on the effect of toxic substances—so "the experimental-pathological concepts" (p. 8)—and on a "change in blood" that "was a popular phrase used to explain all generalized diseases. Whereas it went progressively out of fashion for other diseases, however, its significance only increased in the case of syphilis." (p. 11)

In 1905, a causal agent, Spirochaeta pallida, was found by a team of civil servants under the supervision of German Health Authorities. But the modern definition for syphilis did not appear before the discovery of the Wassermann reaction, a blood test for syphilitic antibodies (amboceptors), found by another team of researchers. Fleck: "They wanted evidence for an antigen or an amboceptors. Instead, they fulfilled the ancient wish of the collective: the demonstration of syphilitic blood." (p. 70)

Fleck continued: "Syphilis is not to be formulated as 'the disease caused by Spirochaeta pallida.' On the contrary, Spirochaeta pallida must be designated 'the micro-organism related to syphilis' " (p. 21). "In the end an edifice of knowledge was erected that nobody had really foreseen or intended. Indeed, it stood in opposition to the anticipation and intentions of the individuals who had helped build it." (p. 69)

The very problem of understanding a fact does not arise within an interpretative framework—or within a thought collective—but rather between such frameworks. How are interpretative frameworks linked one to another? Fleck (1935) showed the differences between popular scientific knowledge and different stages of scientific abstraction, distinguishing:

- *popular science* (e.g., help-yourself medication; everyday knowledge on physics: the space, the Earth, gravitation . . .) that has to grasp common understanding. Therefore, "a vivid picture is created through simplification and valuation" (p. 113);
- *textbook science* (e.g., for students) that organizes scientific knowledge in accordance to conventional scientific theories;
- *vademecum science* (or handbook science) that requires a "critical synopsis in an organized system" (p. 118). Vademecum science is the condensed conventional technical knowledge of a discipline. Its use is limited to professionals and experts of that discipline;

- *journal science* that consists of the research articles widespread in disciplinary journals and that "bears the imprint of the provisional and the personal." (p. 118)

From outside these scientific spheres, there is the need for explanation and interpretation of scientific knowledge for which an expert can serve. No one who has troubles in basic arithmetic would consult a mathematical handbook. It was Fleck's main objective to introduce the influence of thought collectives. In our context, we want to emphasize the necessity of the interpretation of scientific facts. Knowledge is not information. Knowledge is relative to an interpretative framework. From that point of view, scientific experts are knowledge interpreters. Textbooks cannot supply the need for every kind of interpretation. In fact, they need interpretation, too. We can generalize what is said about scientific experts for experts in general. With regard to the variety of spheres of our everyday life (work, communities, generations, hobbies), "everyday" experts, too, can be regarded as knowledge interpreters.

In summary: Experts are knowledge interpreters. They interpret a problem (a question) from the point of view of a sphere of knowledge that is valid within a scientific thought collective or other knowledge communities. The knowledge background of an age determines which interpretations are acceptable. The *Leistung* of an expert (effect and efficiency) for the sake of which experts are used is the relatively fast utilization of the expert's compressed experience any reasonable person could make if he or she had enough time to do so.

CHAPTER

4

In a New Light: Organizational Role Conflicts With Experts, and Their Resolution

In the last chapter, we argued that "The expert" has to be regarded as a social form of interaction, and that "expert" is an attribution to the person who, in this interaction, provides information or explanation, respectively. In short, "expert" exists only in a social context.

In this chapter, we introduce one of the main important social contexts in modern societies: organizations. Organizations set specific constraints on expert services that can result in role conflicts for experts. We present an attributional approach to such role conflicts in experts. In the final part of this chapter, we turn to research in social psychology and see how experts contribute to social validation and the distribution of knowledge.

4.1 INTRODUCTION: ORGANIZATIONS

Organization can take on several forms. It can be a corporation such as a bank or factory, it can be a federal office or an international organization, or it can be an association such as the International Dentists' Federation. Organizations differ from groups in that they—in principle—exist independently of the people who participate. Of course, there would be no International Dentists' Federation without dentists and no factories without workers and managers. Microsoft would probably not be what it is without Bill Gates. However, even Microsoft is not Bill Gates. There are many—we can say nameless—others who work for Microsoft and buy and sell in the name of Microsoft. From a juridical point of view, organizations can be given a form so that they can act as if they were persons, particularly in court. Therefore, we can also speak of organizations as "corporate actors."

Organizations can be regarded from a structural and process-related point of view. An organization's structure comprises such divergent entities as buildings, the firm's legal status, its salary system, the division of labor, or its documented history. The organization's structure determines the conditions for the members of that particular organization. The structure constitutes the conditions for membership. Organizations are social units (systems), and membership denotes the relevant difference between the inside and outside of an organization.

The process-related aspect of an organization deals with the way in which time is structured by that organization. Thus, the process-related aspect comprises working processes, managerial decision making, but also the duration of contracts. From the point of view of the individual member, the process-related aspect is one of role-taking. Although dependent on the structural constraints, any member has to define his or her role. Role-taking is a means for the coordination of the individual behavior of the individual members of the organization. Acting in an organization is a perfect example for what Mead (1924/1925) called a *social action* that is "distributed among a number of individuals" (p. 274). However, the dynamics of an organization are not solely determined by social acts: There are external and internal forces such as the market or the machine times that shape the organizational processes.

There are several approaches toward an organizational theory. We do not introduce all approaches in detail. Instead, we use them to clarify the structural and process-related aspects of organizations.

The Structural View: Bureaucracies

Max Weber introduced the modern concept of a bureaucratic administration. For Weber, bureaucracies are based on the historical process of rationalization. *Rationalization* means that administration and behavior–in the private as well as in the public sphere–are increasingly regulated by impersonal, justifiable rules that are usually laid down in some kind of statutes. Rational authority is ruled by law–therefore, Weber (1947/1964) spoke of legal authority based on the idea that "any given legal norm may be established by agreement or by imposition, on grounds of expediency or rational values or both, with a claim of obedience at least on the part of the members of the corporate group" (p. 329). This is rationality in the sense of impersonality or objectivity. The "purest type" of administration through legal authority is the bureaucratic administrative staff. It consists of individual officials who

- are personally free and subject to authority only with respect to their impersonal, "objective" duties,
- are organized in a clearly defined "hierarchy of offices,"
- exercise clearly defined official competencies,
- are employed by contract and selected according to their technical qualifications (they are appointed, not elected),
- are paid by fixed salaries in money,
- treat their offices as the sole—or at least the primary—occupation,
- can expect a system of "promotion" (career),
- work entirely "separated from ownership of administrative means" and without appropriation of their positions, and
- are subject to a strict, impersonal discipline and control.

Bureaucracies (as pure types) are the perfect realization of impersonal, objective administration. History has shown them to be very efficient, as Weber (1947/1964) claimed:

> Experience tends universally to show that the purely bureaucratic type of administrative organization is, from a purely technical point of view, capable of attaining the highest degree of efficiency and is in this sense formally the most rational known means of carrying out imperative control over human beings. (p. 337)

For Weber, bureaucratic administration means fundamentally: "knowledge-based authority" (see 1947/1964, p. 339; 1972, p. 129).

The Process-Related View: Organizational Decision Making

How does an organization work? March and Simon (1958) studied organizational processes in the framework of organizational problem solving and decision making. Simon (1972) remarked on the theoretical analogy between individual and organizational behavior. The analogy is based on the organizational component that we find in brains as well as in corporations, both being regulated by a procedural rationality.

March and Simon viewed organizational processes in terms of cognitive problem-solving heuristics, particularly the heuristic "factoring a problem into subproblems." What Weber called the *hierarchy of offices* now becomes

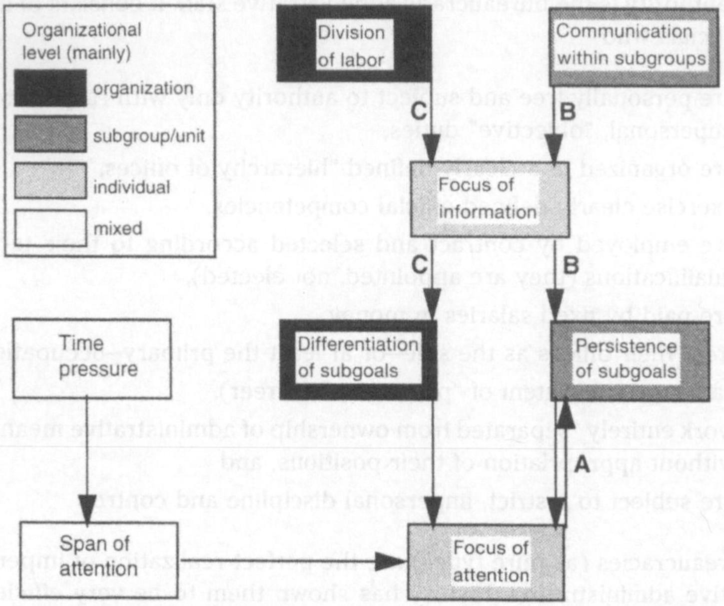

FIG. 4.1. Some factors that affect "selective attention to subgoals" (according to March & Simon, 1958/1993). Time pressure usually reduces the span of attention. Therefore, the focus of attention is reduced, too. From the point of view of individual cognition, the span of attention is the capacity of short-term memory. *Note.* From *Organizations* (Fig. 6.1) by J. G. March and H. A. Simon, 1958/1993, Cambridge, MA: Blackwell. Copyright 1958/1993 by Blackwell. Adapted by permission.

a system of subgoals into which we can factorize the goals of the organization. March and Simon distinguished three main organizational levels: the organization as a whole, the subgroups or units within the organization, and the individuals. March and Simon (1958/1993) explained, "The principal way to factor a problem is to construct a means–end analysis. The means that are specified in this way become subgoals which may be assigned to individual organizational units" (p. 173).

The organizational factoring of goals into subgoals or organizational units, respectively, may result in other subgoals and other aspects of the goals of the larger organization being ignored. Figure 4.1 shows March and Simon's analysis of the reinforcement of the "tendency of members of organizational units to evaluate action only in terms of subgoals" (="persistence of subgoals"). For instance, the marketing division is primarily concerned with producing good marketing (=subgoal) and only indirectly translates into action the "mission" of the organization (=goal) or the overall goal of realizing profit. The same is true for other divisions. March and Simon stated sources of reinforcement at each organizational level:

- at the individual level, there is reinforcement through "selective perception and rationalization." The persistence of subgoals is reinforced by the "focus of attention" on that subgoal (Path A).
- at the subgroup level, it is the type of communication within subgroup that affects the focus of information within the subgroup and reinforces the persistence of the subgoal (Path B).
- at the organizational level, it is the division of labor that affects the selection of information that members of the organizational units receive and that reinforces the specific differentiation of subgoals (Path C). Reinforcing the specific differentiation of subgoals at the level of the organization means persistence of subgoals at subgroup level.

We have already introduced both organizational aspects—the structural and the process-related one—in a rather idealized way. Neither Weber nor March and Simon believed that all or most organizations behave in such a rational way. March and Simon explicitly discussed organizations under the constraints of "cognitive limits on rationality." As the French sociologist Crozier (1964) remarked, the reality of bureaucratic administration is far from being perfect: "Reality [in organizations, HAM], in fact, appears extremely different from the perfect administration of things" (p. 158).

In his book *The Bureaucratic Phenomenon*, Crozier (1963/1964) analyzed the administrative staff in several corporations. He found evidence for phenomena that did not fit into the picture of an impersonal bureaucratic staff that tries to solve problems. In particular, he found competition among administrative units. Moreover, sometimes bureaucracies reinforce or invent problems to justify their existence and budget. Crozier analyzed the phenomenon in terms of power: Bureaucracies as well as subordinates and managers compete for influence with each other. The reason is: There is no best solution for many problems that administrations are concerned with. Faced with uncertainties, it is also up to the administration to define the problem space. We can say: Bureaucracies often compete for the power of defining the relevant problems.

Another phenomenon in organizations that is often neglected is *time*. The British psychologist Elliott Jaques (1976) studied how time is organized in corporations and how individuals cope with the organization of time. He found that time in corporations is organized according to levels of abstraction. At a low level of abstraction, more concrete work is organized in a rather limited time-span:

> It is a general feature of roles with time-spans below 3 months that the tasks are assigned in concrete terms and are carried out in direct physical contact with the output [. . .] For example; a copy-typist is given a typed memorandum to copy, and works physically with her machine in doing so; a manual operator is given a drawing from which to carry out a particular operation on a piece of metal either by hand and or by a machine [. . .] (p. 144)

There is a profound change in the quality of work for time-spans longer than 2 years: a change "from the concrete to the abstract mode of thought and work" (p. 147). For this time-span, we find conceptual or managerial types of work. Conceptual work requires abstraction in the sense of detachment—or "being able to work at specific and concrete problems without dependence upon mental contact with existing things" (p. 150). Thus, in organizations we find not only the allocation of subgoals to organizational units, but also the selection of individuals according to their ability to work with different levels of time or abstraction, respectively:

> The capacity to manage activity through time is the counterpart of, and depends upon, the capacity to analyze and detail situations, to pattern and order the detail. The capacity to analyze, pattern and order detail depends upon the organizing and conceptualizing capacity of the mind. (p. 157)

To conclude, we redefine organization. The definition should take into account, or be open to, the various organizational aspects we have discussed: impersonality of bureaucratic administrations; factoring of organizational goals; uncertainty, and intraorganizational competition; and time organization and selection of individuals. We can say this: Organizations are resource pools (see Coleman, 1974). For a certain time, individuals place parts of their resources at the disposal of a central unit that is not identical to them. Those resources may be work or money and are used to attain organizational goals. There is competition for resources—including experts. Therefore, managing organizations is also a problem of allocating resources.

4.2 EXPERTS' AUTONOMY

In theory and practice, experts encounter role conflicts in organizations. The reason is the autonomy of expert work. Literature on professionals and experts sometimes refers to the autonomy of experts as one of the core features of the expert's role. Autonomy means self-regulation. The "rules of the game" the experts play are defined by themselves. For example, medicine as a profession defines the criteria both for the education of physicians and the standards of professional medical work (see Freidson, 1970a, 1970b). In principle, autonomy seems to be in conflict with organizational constraints. If organizations solve problems in the manner discussed, the definition of the problem space is the obligation of the management. The subgroups only work on subproblems or toward subgoals. Thus, professionals (and experts) within the staff suffer—in theory—from being left with a reduced base for diagnosis or knowledge interpretation. We discuss the issue with regard to the concepts for expert services we called *function*, *status*, and *person* in chapter 3.

Function

There seem to be functional role conflicts of professionals in organizations. In 1968, Richard H. Hall published a study on professionalization and bureaucratization that showed functional conflicts of professionals in organizations. The study differentiated between a structural and an attitudinal dimension of professional autonomy:

> While the structural aspect of autonomy is directly subsumed under the efforts of professional associations to exclude the unqualified and to provide for the legal right to practice, autonomy is also part of the setting wherein the professional is expected to utilize his judgment and will expect that only other professionals will be competent to question his judgment. The autonomy attribute also contains an attitudinal dimension: the belief of the professional that he is free to exercise this type of judgment and decision making. (p. 93)

In contrast, the degree of bureaucratization was defined in accordance with Max Weber's definition of *bureaucratic administration*. Hall investigated professional work within organizations (e.g., physicians in hospitals, lawyers in law firms and law departments of an organization, social workers). He found a negative correlation among almost all of his criteria for bureaucratization and the feeling of autonomy. Bureaucratization and the attitudinal dimension of autonomy seem to be mutually exclusive. Moreover, members of professions with high structural autonomy (physicians, lawyers) worked in organizations with less bureaucratic structure than members of professions with low structural autonomy (nurses, accountants in accounting departments).

However, 30 years of research on role conflicts of professionals within bureaucratic organizations have not yielded consistent results. For example, studies on accountants in professional certified public accountant (CPA) firms report both severe role conflicts (Sorensen & Sorensen, 1974) and an overall low level of organizational-professional conflict (Aranya, Pollack, & Amernic, 1981). Also, studies on corporate law departments (lawyers are high-autonomy professionals) and accountants (low-autonomy professionals) report similar job satisfaction (Aranya & Ferris, 1984; Gunz & Gunz, 1994). Moreover, commitment of employed lawyers to their corporation does not seem to exclude commitment to their profession (Gunz & Gunz, 1994)—a double commitment that was found to be impossible for nurses (Corwin, 1961). Thus, there seem to be role conflicts, but they do not necessarily surface when professionals work in organizations.

Status

Regarding the expert's status, autonomy can be considered as a specific privilege. It is the privilege of the status of a third party (see Hitzler, 1994). In this sense, autonomy gives the impression of neutrality and objectivity. This is the reason that we find experts who serve to legitimize specific interests (e.g., the interest of a political party or pressure group that try to make believe that its particular interest is supported by scientific knowledge). As a matter of fact, we find a variety of contexts in which experts help promote products, persons, or decisions—from daily advertising to international politics.

From this point of view, the experts' autonomous status is a claim made versus a certain audience. We can distinguish between large audiences—the mass media public—and small audiences—a limited public. Depending on the audience, the use of experts differs. Table 4.1 shows common sources for legitimizing uses of experts. In the political domain, we also find the administrations using expert panels to demonstrate the severity of the problems and the necessity of administrative work, thus creating a demand for bureaucratic staff and funds, as Jasanoff (1994) described it for the relation between Environmental Protection Agency (EPA) and its Science Advisory Board.

TABLE 4.1
Sources for Legitimizing Uses of Experts

	Kind of Public	
Domain	*Mass Media Public*	*Limited Public*
Politics	Parties, pressure groups	Administrations
Corporations	Advertising, PR	Consulting
Law	Expert witnesses (U.S. law)	Judge–expert relationship
Press	"Experts" columns	Investigative journalism
Science	Research funds	Peer reviewing
Professions	Image, licenses	Client relationship

Note. Only distinctive domains and uses of experts are mentioned. The table distinguishes between scientists and professionals, a distinction we do not want to emphasize. Many a profession comprises science and business (e.g., medicine) because each fully fledged profession seems to need an academic basis. In contrast, there are sciences without professional business outside university (e.g., Ancient Greek).

Corporations and other organizations use experts not only for advertising and general public relation purposes, but also as consultants to the management. Even when a decision has been made, it can be wise to bring in a consultant who supports the decision. Support by McKinsey or another top consulting company justifies the measures to be taken. According to his biographer, James O. McKinsey "decided that for a successful consulting practice he needed three ingredients":

- "unquestioned respectability";
- "a reputation for expertise in an area of some concern to management"; and
- "professional exposure." (Neukom, 1975, p. 3)

These three ingredients also describe the prerequisites of the role of a consultant in general.

Table 4.1 also includes experts in courts and in relation to the press. Here the interaction can be restricted to two persons: a court-appointed expert informing a judge or an expert providing background knowledge for a journalist or an author; the interaction between scientists and journalists is also open to conflicts and mutual misunderstanding (Peters, 1995). Science and professions, as referred to in the last two rows, differ from the other domains in that they usually supply experts. For their part, too, special interests can be put forward, especially for governmental or public research funds and, in the case of professions, specific market regulations that exclude other professions. Within science, we find systems of peer review: Articles in purely scientific journals usually have to be evaluated by other scientists, thus functioning as experts within science. The most exclusive source for a legitimizing use of experts appears in the relationship between an individual professional and the client (e.g., a doctor and a patient). The autonomous expert status enables a professional to legitimize—from a professional point of view—his or her own decisions and therapies.

Person

Autonomy of experts can also be viewed from a psychological point of view. Lawyers and judges might have to ask themselves: Who is a real expert? The psychology of expertise—until now—has had no problems when it comes to identifying experts. Robert R. Hoffman, in his recent summary on knowledge elicitation, resumed the source for selecting experts:

> "Experts" have been selected on the basis of professional criteria (graduate degrees, training experience, publication record, memberships in professional societies, licensing, etc.). . . . Experts have been selected by virtue of the fact that they held down jobs in operational settings, and by the simple process of

> asking workers to identify the expert within their organizations. (Hoffman, Shadbolt, Burton, & Klein, 1995, p. 131)

Psychologists “relied on the participation of an avid dinosaur fan, a 4-year old child,” as well as on “the participation of preschool children who were avid fans of ‘Star Wars’ films” (p. 131).

In general, the psychology of expertise is based on a developmental concept, stretching from the layperson, over the novice to the expert who is “highly regarded by peers, whose judgments are uncommonly accurate and reliable, whose performance shows consummate skills and economy of effort, and who can deal effectively with rare or ‘tough’ cases” (Hoffman et al., 1995, p. 132, Table 1). Thus, the summit of becoming an expert can be seen in the “master,” the one “who is qualified to teach those at lower levels” (loc. cit.). Unintentionally or not, the development is described in analogy to a monastic order. The monastic order is a metaphor for experts that is widely in use in commonsense psychology. The *monastic order metaphor* grasps the idea of truth as a value: Laypersons do not know the (scientific, professional) truth the master knows or believes in, respectively. It is unnecessary to point out that a monastic order does not seem appropriate to describe roles in modern societies. Moreover, the monastic order metaphor neglects the social dynamics of the expert’s roles and tends to identify expert roles with experienced persons. We see that this is one source for role conflict for experts.

4.3 AN ATTRIBUTIONAL APPROACH TO ROLE CONFLICT FOR EXPERTS

Imagine the following situation: You are a biologist working part time for a national research center for groundwater and part time in your own company. You earn a large part of your living by providing expert opinion and reports for national offices. You are asked for your expert opinion on a case of groundwater contamination through a waste disposal site. The customer—the local office for water protection, your main customer—urges you to find evidence for contamination so that it can start to renovate the waste disposal site. However, you cannot find any evidence for groundwater contamination.

Traditionally, the role conflict in this scene would be analyzed in terms of professional autonomy that comes into conflict with bureaucratic constraints. We show that there are several sources for potential role conflicts in this scenario, and that these sources are to be found in the manner in which we attribute the expert role to a person.

We said that there are three attributional aspects: the “expert” attribution can be personal, dispositional, and causal. None of these attributions needs to be valid. Nonetheless, each can cause conflicts. Figure 4.2 shows the aspects of the attribution involved in the social form "The expert": A person is addressed as an expert by/in front of an audience.

Addressing someone as an "expert" can result in conflicts for the simple reason that the person addressed by us might not think of him or herself as an expert in that situation. Let us assume a person has skin problems and seeks the advice of her friend who is a physician; then the doctor (addressed as "expert") might check:

- her relative expertise: She will define her expertise depending on whether the patient is a doctor, too, or whether other presumably competent doctors are around;
- her personal (objective) expertise: How much does she know about skin diseases?
- the relevance of the case: Is it an allergy? Are there dermatological tests to be conducted? Is it a disease at all?

Then the physician might provide an explanation by interpreting her knowledge in this particular situation. The explanation depends on the problem, the medical knowledge, and, of course, the knowledge of the nonexpert. The doctor, seeing this particular skin disease, might find herself incompetent (lacking competence), or she knows that it is a new and rare kind of

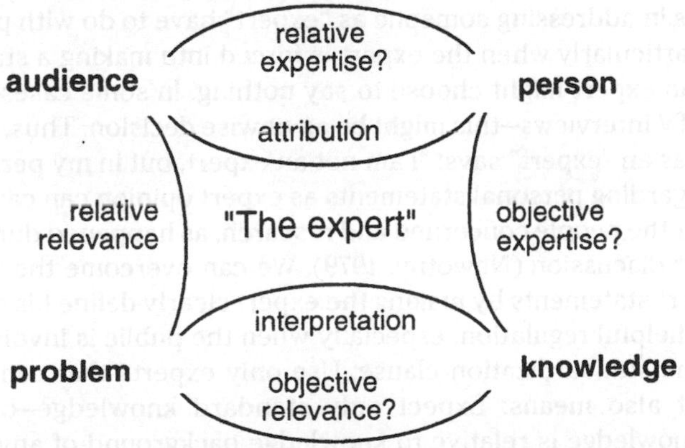

FIG. 4.2. Aspects of the social form "The expert." A person is addressed as expert by/in front of an audience. Generally speaking, the *Leistung* of the expert is to find an interpretation for a problem within a knowledge framework. The person addressed as "expert" might question his or her own relative competence (in comparison to the audience); his or her objective expertise; and the relevance of the knowledge for solving/interpreting the problem. The figure does not include personal motives of the experts or the audience (e.g., animosity).

TABLE 4.2
Sources of Role Conflict for Experts

Aspect of Attribution	*Potential Conflict*	*Possible Regulation*
Personal	Forced statement ("An expert has to know.")	Defining competence and relevance / *standardization clause*
Dispositional	Expecting general knowledge ("Experts know a lot.")	Narrowing the question (in this particular situation) / *generalization clause*
Causal	Forcing decisions ("In virtue of their knowledge experts should actually decide.")	Separating decision base and actual decision / *information clause*

skin disease for which no specific treatment has been developed yet (lacking objective relevance). Or the doctor may decide not to give any specific explanation in this particular case. Perhaps the person asking is a child and, for the time being, the best therapy seems not to worry about the skin. In this case, the doctor also considers the relative relevance of the problem for the patient. Thus, there might be many reasons for the refusal of an "expert" role. Table 4.2 shows all aspects of attribution and the potential conflicts involved.

Forced Statement

Mismatches in addressing someone as "expert" have to do with personal attribution particularly when the expert is forced into making a statement. In that case, an expert might choose to say nothing. In some cases—for example, in live TV interviews—this might be an unwise decision. Thus, the person addressed as an "expert" says: "I am not an expert, but in my personal opinion. . . ." Regarding personal statements as expert opinion can cause difficulties both for the people concerned and research, as happened during the Nuclear Power discussion (Nowotny, 1979). We can overcome the problem of forced expert statements by making the expert clearly define his or her competence. A helpful regulation, especially when the public is involved, is provided by the standardization clause: Use only experts that can be substituted. That also means: Expect only standard knowledge—once again, standard knowledge is relative to knowledge background of any particular given time.

Expecting General Knowledge

Another type of conflict is caused by expecting a too high degree of general knowledge from the expert. *Generalization* is a fundamental principle of learning and cognition. Dispositional attributions are generalizations: Someone who talks about laws and cases might be regarded by an audience as possessing judicial expertise. Then by virtue of this expertise, the person seems to be able to talk about laws and cases in other contexts, too. Thus, dispositional attributions generalize over time and place. The generalizations become problematic if they reach into other domains of knowledge. We might believe that an excellent organizational psychologist might be quite adept at providing psychotherapy, that a person who knows a lot about flowers also knows a great deal about gardening, and so on. As already discussed, expertise normally is domain-specific, one reason being that sound expertise takes about 10 years to develop.

Another more subtle type of generalization concerns types of information and explanations an expert can provide. As Table 4.3 shows, there are several types of explanations that are based on different kinds of knowledge and that necessitate different kinds of validating criteria. What-explanations answer what-questions (What is syphilis? At what age did Roosevelt become president?) and refer to facts. In this case, explanation is some sort of information. The statements about facts can be true or not. As we have discussed in the case of syphilis, a fact is relative to the knowledge background of any particular time. From the point of view of cognitive psychology, what-explanations refer to declarative knowledge such as we find in textbooks.

How-explanations refer to procedures (How do I play an endgame in chess? How do you cure allergies?). Procedures are measured by their effectiveness (Did the therapy help?). As discussed earlier, we can describe increasing expertise as the transformation of declarative knowledge into procedural knowledge (Anderson, 1985). It has also been discovered that where the experts' procedures are based on routine, we might face difficulties in getting valid information from experts on how they actually proceed (Nisbett & DeCamp Wilson, 1977).

Why-explanations refer to causes and reasons. Causes are linked to general laws (e.g., physical laws), which explain why certain phenomena appear (Why does water freeze? Why do prices increase with demand?). Reason refers to metaphysical principles that justify our belief in certain laws (Why is there life at all?). From a psychological point of view, why-

TABLE 4.3
Type of Explanation Provided by Experts

Type of Explanation	*Type of Knowledge*	*Validity Criteria*	*Epistemic Category*
What	Facts	True	Declarative knowledge
How	Procedures	Effective	Procedural knowledge
Why	Cause and reason	Reasonable	Metacognitive knowledge

explanations are based on some kind of monitoring or metacognitive knowledge (Flavell & Wellman, 1977; Metcalfe, 1993) that helps control and coordinate declarative and procedural knowledge. Switching between categories of explanation can hinder interaction with experts. A historian might refuse why-questions as speculations and stick to facts. Similarly, a cook might not be inclined to really explain any causes or reasons for cooking.

Regarding interactional problems caused by undue generalizations, it is useful to narrow the specific question in accordance with Fig. 4.2; when several experts are present, they usually define their competence in contrast to one another. Another useful strategy that has already been introduced is the generalization clause. The generalization clause formalizes the output of the expert's *Leistung*. It states that every reasonable person should be able to follow the logical chain that led to the explanations. Then unbiased inferences or undue generalizations can be detected more easily.

Forcing Decisions

One of the most common conflicts the social form "The expert" can bring about for experts, especially nonprofessional experts, is forcing them into making a decision. Usually experts do not decide—they explain.

> Here we have to make a note on the theory of rational decision making. Rational decision making means choosing between alternatives (e.g., Luce & Raiffa, 1957). The theory of decision making can be used in a descriptive manner and for prescriptive purposes. We can describe judgment and problem solving in terms of decision making, since in problem solving we have to choose between possible ways for resolution and judging can be seen as the problem of finding the suitable judgment (among several possible judgments). Then, by definition, experts decide: They decide on what is the correct answer. (facts, procedure, causes, reasons)

The provision of explanations is the specific *Leistung* of experts. Demanding that an expert should also decide transcends the social form "The expert" and requires another kind of contract (e.g., a professional contract). Professionals (e.g., lawyers or physicians), even if the ultimate discretion is on the part of the clients, decide on the fate of their clients, as far as their decisions are justified by professional standards. The physician decides how to apply medication or how to carry out an operation, and the lawyer decides how to sue the adversary and realize the client's claims. However, asking somebody on the street for directions to the station does not imply that the person asked has to take us to the station or that this person is really expected to decide what way we will take. Usually the person who advises and the one who decides are not the same. Experts can face role conflicts if they are urged to decide. This was the case at a Swiss office that tried to use experts to support a project "of common interest," but lacked scientific necessity:

> A Swiss cantonal office had to decide on a project for the redevelopment of a waste deposit site. The waste deposit site contained barrels with toxic PCB and other liquid toxins; it was situated on a mountain slope in the neighborhood of an important groundwater reservoir. The initiative was on the part of the office: but the immense projected costs (several million Swiss francs) required political support. To this end, an expert panel was installed (one geologist, one biologist, one engineer, all specialized in waste disposal), chaired by a government employed environmental specialist. The panel's task was to evaluate the risk the waste deposit site represented to the groundwater. But, several scientific tests brought neither evidence for a leak nor for movement of the mountain slope. The biologist admitted:
>
> > "I would really like to testify as to the risk of that deposit, but looking at the scientific data I cannot."
>
> Nevertheless, the office clearly demonstrated that they wanted the evidence. Then the chairman introduced a plan for opening some of the—fairly smelly—barrels in presence of the press. The experts were asked to support the plan. The experts vehemently refused. On the whole, they felt misused and, as a consequence, reduced communication on their part.

We can say: Experts help define the decision base; the actual decision is up to someone else. This is the core idea of the information clause that restricts the expert's part to information and instruction. We can say that it would mean wasting human expertise if experts were generally accountable for the decisions that are based on their information, instruction, or explanation.

To summarize: Role conflicts in experts do not only occur in organizations and are not always linked to conflicts with the experts' autonomy. Instead, we can analyze role conflicts in experts by reference to the attributional aspects in the "The expert"-interaction (as defined in chap. 3.2). Moreover, the principles we found for the use of experts in judicial processes—according to the Continental law tradition—may help prevent such role conflicts.

4.4 SOCIAL PSYCHOLOGY OF EXPERT ROLES

From the point of view of social psychology, the expert's role can be characterized by both the specific role expectations or attributions, respectively, and the specific *Leistung* that is the use of knowledge by experts. We discussed the conflicts that role expectations cause for experts, particularly in organizations. We now turn our attention to the *Leistung* aspect and ask: How does the expert's role contribute to forms of "collective cognition" such as organizational problem solving? Mead (1934) conceived of the mind as a social phenomenon. According to Mead, mind is a product of the need for social coordination. Modern social psychology has revealed some mechanisms of collective cognition within experts. In particular, we want to draw attention to Wegner's (1987) work on transactive memory and the work of Garold Stasser on expert roles and the use of unshared information.

Transactive memory means the coordinated and distributed storage of knowledge within groups. To explicate the phenomenon of transactive memory, Wegner (1987) focused on the use of external storage for many everyday memory tasks. We use notebooks and agendas to remember appointments or events. We take notes in seminars, we prepare shopping lists, and many a speaker uses manuscripts when preparing a speech. Moreover, books, files, and microfilms store a large part of the cultural and professional knowledge that societies have accumulated to date. We might also use other people as external memory—for instance, by asking a sister or brother to remind us in time of our parents' birthdays. A large part of a secretary's job consists of reminding her superiors of appointments.

Memory processes are regulated by metacognitive knowledge that is knowledge about how to retrieve specific memory entries. This is true for internal as well as external forms of memory. If we try to remember the name or face of a person, it might be helpful to think of the occasions when we met this particular person. It might also be helpful to look in our address book or at an album with photographs. Analogously, we use dictionaries and encyclopedias to locate information that we do not remember. In this case, metacognitive knowledge means: Even if we do not know a specific fact, we know how to gain access to information about it. The same is true if we use other people as external memory. Wegner (1987) stated, "[O]ne person has access to information in another's memory by virtue of knowing that the other person is a location for an item with a certain label" (p. 189). Thus, transactive memory is "a property of a group" (p. 191)—for instance, a family or an organization.

A central mechanism in a transactive memory is the attribution of expertise. In a laboratory memory experiment, Guiliano and Wegner (1985) showed the effects of attributing expert roles in intimate couples. The couples were asked to memorize certain items together. As Table 4.4 shows, best recall was found in persons who, on the one hand, were "experts" and responsible for the particular piece of information and, on the other hand, whose partners were nonexperts.

The phenomenon of transactive memory demonstrates the relativity of the attributed expertise. Attributed expertise has to be seen in the context of the "The expert"-interaction and the problem to be solved (as Fig. 4.2 explicated). In couples, transactive memory strategies ensure "that informa-

TABLE 4.4
Percentage of Recall as a Function of Attributed Expertise and Situational Responsibility for Knowledge in a Memory Experiment With 20 Intimate Couples

Who Is Responsible for Knowledge in the Particular Step of the Experiment?		*Self Is Expert (%)*	*Self Is Nonexpert (%)*
Self	Partner is nonexpert	34.3	29.8
Self	Partner is expert	27.3	24.2
Partner	Partner is nonexpert	24.9	17.3
Partner	Partner is expert	23.6	19.8

Note. Data are from "Transactive Memory: A Contemporary Analysis of the Group Mind" by D. M. Wegner, in *Theories of Group Behavior* (Table 9.1) by B. Mullen and G. R. Goethals, 1987, New York: Springer. Copyright 1973 by Springer. Reprinted by permission. Recall is highest when the self is expert and responsible and the partner is not; recall is lowest when the self is nonexpert and the partner is responsible for knowledge in that situation. Wegner (1987, Table 9.1) reported: When the self is in this situation responsible, the differences in recall according to the partner's expertise (expert vs. nonexpert) are statistically significant (analysis of variances, $p < .01$). When the partner is in this situation responsible, the differences in recall according to the self's expertise (expert vs. nonexpert) are statistically significant ($p < .05$).

tion the couple needs will always be captured by at least one of the partners" (Wegner, 1987, p. 194). A transactive memory develops because responsibility for knowledge is distributed within a group:

> The construction of a working transactive memory in a group is a fairly automatic consequence of social perception. We each attend to what others are like and in this enterprise learn as well what we can expect them to know. Then, when the group is called upon to remember something, information is channeled to the known experts. When no expert is known to exist, the individual who is entrusted with the information by circumstance holds on to it, allowing the group subsequent access. In sum, transactive memory can be built because individuals in a group accept responsibility for knowledge. (p. 194)

From the point of view of transactive memory, the "selective attention to subgoals" in organizations, as examined by March and Simon (1958/1993; see Fig. 3.4), is simply the byproduct of a developed organizational transactive memory system.

Not every assembly of persons automatically becomes a transactive memory, not to mention a team. Even in case of work groups or teams, collective performance is not automatically superior to individual performance. The relative performance of individuals and groups, particularly of experimental ad hoc groups, heavily depends on the task. In an overview,

Levine, Resnick, and Higgins (1993) concluded that the critical determinant is *solution demonstrability*: "['S]olution demonstrability' is the critical determinant of a group's ability to develop an adequate shared representation, with groups performing best when the task has a correct solution that can be readily demonstrated and communicated to members" (pp. 600–601).

Garold Stasser and colleagues investigated conditions for the effective use of knowledge in groups. Their focus was on so-called *hidden profiles* in experimental problem-solving groups. In a hidden profile, a superior solution exists, but its superiority is hidden from individual members of the group because each individual has only one portion of the information that supports the superior solution (Stasser 1992a, 1992b). Hidden profiles can be revealed if the group members pool their unshared information. When faced with a hidden profile, the group can find a solution or decision that is superior to potential solutions by the individual group members. In general, there is a tendency in groups to omit unshared information (Stasser & Titus, 1985). A more recent study by Stasser and Stewart (1992) showed that groups are more likely to reveal a hidden profile when they assume that the experimental task has "a demonstrable correct answer" (p. 434). As we see, however, the most effective strategy seems to clearly assign expert roles for unshared information.

In an experimental study on hidden profiles, Stewart and Stasser (1995) investigated the use of unshared information and expert role assignment in decision-making groups. In this study, unshared information was a basis for the definition of expertise, expertise signifying "that a person has access to more information in a specific domain than others in the group" (p. 619). The study was composed of three-person groups that evaluated applications, with some members—the experts—having special valuable information on the applicants. In some groups, the expert role was assigned; in others, it was not. The groups had to prepare reports. The study investigated how much unshared information appeared in the reports. As Table 4.5 shows, the amount of unshared information in reports increased when the expert role was assigned. In the "no-assignment" condition, the amount was 16% in comparison to 41% in the expertise-assigned groups.

The study concluded that the assignment of expert roles serves as a source of social validation—that is, "the veracity of the information introduced by one group member is confirmed by another" (p. 627). From this and other results on expert role assignment, Stasser and colleagues inferred conditions for effective expert role assignment in groups (see Stasser, Stewart, & Wittenbaum, 1995):

- the members expertise status needs "to be explicitly recognized within the group";

TABLE 4.5
Social Validation

Expertise	*Unshared Information*	*Shared Information*
Not assigned	16%	47%
Assigned	41%	51%

Note. From "Expert Role Assignment and Information Sampling During Collective Recall and Decision Making" by D. D. Stewart and G. Stasser, 1995, *Journal of Personality and Social Psychology, 69*, p. 625. Copyright 1995 by American Psychological Association. Reprinted by permission. In groups where expert roles are assigned, more unshared expert information is taken into account than in groups where no expert roles are assigned (i.e., some individuals had valuable unshared information but were not assigned as "experts").

- members need to be "mutually aware of each other's area of expertise at the onset of discussion to facilitate dissemination of unshared information";
- expert role assignments need to fit the actual distribution of knowledge:

"Divisions of labor based on role assignments that do not correspond to access to unshared information could easily aggravate the tendency to omit unshared items from discussion." (p. 263)

In conclusion, we return to the "value of truth" that accompanies "The expert"-interaction. We can say that the "value of truth" is nothing more than the profit of social validation in cases where we assume that a superior solution or a correct answer is possible. Experts are persons used as knowledge, this knowledge being true insofar as we—or the "generalized other"—can gain knowledge in this case.

CHAPTER

5

Case Study I: Experts–Risk–Financial Markets

The first case study is on a case of strong public interest: experts and financial markets, particularly stock markets. U.S. legal decisions early in the 20th century made expert advice necessary for many trustees—for example, trustees of pension funds (see Burk, 1988). This gave rise to the stockbroking profession.

In Shanteau's performance listing of experts (see Table 2.3), stockbrokers are listed on the side of the poor performers. This is an astonishing fact for several reasons:

(a) There seems to be a clear idea about performance in stock markets: Gains and losses are effectively measured in price differences; each participant can be clearly and immediately informed about every stock price; all of the data on previous prices and market changes are accessible. Thus, the stock market seems a paradigmatic case for rational decision making.

(b) Stock markets are within the focus of an encompassing and still growing body of economical research and theory. We introduce Markowitz's (1952, 1959) theory of portfolio selection; but there is even more theory and research on markets, finance, and investments that has reached the status of textbook science (e.g., Sharpe, Alexander, & Bailey, 1995), notwithstanding the fully fledged professions of financial business.

One key to understanding expertise in financial markets, from a theoretical as well as a practical point of view, is human behavior toward risk—that is, assessing, predicting, and managing risks. The first section prepares the ground for the case study, displaying two markedly different approaches for understanding risk: a more formal one—risk as volatility—and a more sociological one—risk as commitment. They describe human risk behavior equally well, but come to diverging conclusions as to standards of dealing with risk.

The second section introduces the domain of finance and stock markets and investigates the specific role of experts. In a nutshell, we find the phenomenon of *blind experts*, who are successively selected due to the always limited value of their particular workable hypotheses and strategies. Moreover, experts' advice seems to enhance volatility—that is, an increase in risk.

5.1 TWO CONCEPTS OF RISK: VARIANCE VERSUS COMMITMENT

Even scholars of risk analysis might wonder, despite some formal consensus, just how divergent concepts of risk are. To understand what expertise in dealing with risks could mean, we have to go into the details of these risk concepts. We discuss two lines of conceptualization. The first, more formal one has its origin in the theory of rational decision making and leads us to the portfolio theory by Markowitz. He considered risk as *variance*. For the second concept, we start with the psychology of risk perception and come to Luhmann's sociological concept of risk that considers risk in terms of *commitment*.

The plan here is to center discussion on the two outstanding scholars—Markowitz and Luhmann—starting each time with some basics of risk analysis:

- first, elements of rational decision making (introducing Markowitz' concept),
- second, elements of risk perception (introducing Luhmann's concept).

This section concludes with a synopsis, where we reveal the different fundamental assumptions underlying the two approaches: on the one hand, the as-if-we-all-knew assumption that everyone equally has access to all the necessary information (relevant in case of Markowitz); on the other hand, the time-and-commitment assumption that there is no decision without the commitment of an accountable concrete decision-making subject such as a person or a firm (relevant in the case of Luhmann). The difference seems minor at first glance, but results in diverging views when it comes to interpreting financial markets.

Some Basics of the Risk Concept From the Point of View of Rational Decision Making

Despite some discord in the details, a framework for a formal understanding of risk has developed in the last 50 years. One point of disagreement is the importance of loss. Does risk always imply a kind of loss? Or is *risk* just another word for *uncertainty*? Let us assume we want to buy a new car. Buying a particular car has different consequences–positive ones such as more flexibility or a certain prestige embodied by the car, and negative ones such as price or maintenance costs. Some of the consequences are more or less certain (the price), whereas others are not (gaining flexibility). The consequences differ with each car. If we cannot decide between two cars, shall we call this situation risky because of the uncertainty of the consequences? Or must there be some possible losses (e.g., heavy repair costs) when we buy a used car? We leave the decision on this point open. The framework for a formal understanding of risk contains these elements (Vlek & Stallen, 1980; Yates & Stone, 1992):

- the possibility of loss,
- a valuation of the possible consequences (losses), and
- uncertainty of the consequences (probability of loss).

Losses mean not necessarily lost money. We can also lose reputation, time, quality, and efficiency of work, safety, love, and so on. With regard to possible losses, we can define *risk* in several ways; risk can be:

- the *worst credible case*. If there is a credible chance that the particular used car, which we intend to buy, will completely break down within the next week, we might consider this the definite worst case and refrain from buying it. Risk as the worst credible case plays a role in the public discussion on environmental issues–for instance, nuclear power plants or waste deposits.
- the *probability of loss*. If we know from a consumer magazine that the new car we are going to buy is from a series that tends to have a high instance of breakdowns and excessive maintenance costs, this piece of information might make us refrain from buying the car.
- the *expected overall loss*. We can also conceive of risk as a mathematical random variable. This requires that there is a (a) defined space of events (e.g., all possible consequences from buying a particular car, such as breakdown, insurance fees, etc.), (b) measurement for all these events and their combination (e.g., money), and (c) distribution for these events (e.g., a probability for each event). If we conceive of risk in this way, we make two strong assumptions for risk–namely:

(a) *multiplicative interaction,* because we multiply risk probability and risk measure

Risk (break down) = probability (of break down) *
measure (in form of repair costs)

This means, for instance, that an increase in risk probability (e.g., doubling from 1% to 2%) equals a similar increase in the risk measure (e.g., doubling from $1500 to $3000), since the overall multiplicative result is the same (1% * 3000 $ = 2 % * 1500 $ = 30). But, it might be difficult to equate the risk of a low probability–high damage event (e.g., accident caused by engine failure) and a high probability–low damage event (e.g., engine noises).

(b) *independence and cumulativity,* because we have to add all risks:

Overall risk (bought car) = Risk (break down) + Risk (high insurance)
+ . . . + Risk (car is attractive to thieves).

The fundamental approach to risk from the point of rational decision making is the calculation of expected utility. The basic idea of the expected utility concept (as introduced in chap. 2, this volume) is to provide one unifying measurement: utility. However diverse things such as money, travel, health, marriages, or car accidents are, they all can be measured and compared by their utility. *Utility* means relative preference: If we prefer a vacation in Paris to buying a used car, the utility of the trip is greater than the utility of buying car:

u (trip to Paris) > u (buying a used car)

Moreover, the utility measure also says how much we prefer one to the other, for instance:

u (travel to Paris) = 2 * u (buying a used car)

Expected utility means the result of combining utility and probability of a decision, for instance: Although the utility of traveling to Paris is high, this trip might have a low probability (e.g., 10% because of a sick relative we have to care for at home). Therefore, the expected utility of buying a used car might outweigh the expected utility of traveling to Paris.

The expected utility approach provides a powerful framework especially in economics. Given the comparability of things—commodities, services—we can estimate rational decision strategies. As to risk, we can search for an adequate compensation of losses. The probability that our home burns down is low but reasonable. In case of a fire, however, the necessary investments would be high. Thus, we can insure our home by paying a certain small amount (a loss) in exchange for the compensation of an uncertain high loss.

There are some restrictions. The expected utility approach not only presupposes general comparability (and, of course, compensability), but also, for example, transitivity of preferences. Transitivity means:

> if decision A (e.g., receiving a $100,000 salary) is preferred to a decision B (e.g., receiving a $60,000 salary), and B is preferred to C (e.g., receiving a $20,000 salary), then we should also prefer A to C.

Transitivity is not valid in cases where, for example, someone cannot decide on the preferred color of the car he wants to buy in such a way that: He prefers blue to red, but red to white and white to blue. But in this case, he would not come to a rational decision either (as to these three colors).

Daniel Kahneman and Amos Tversky criticized the expected utility approach. It cannot explain the peculiarities of human decision making under risk. In 1979, Kahneman and Tversky introduced a revised model—called *prospect theory*—on the following assumptions (see also Fig. 5.1):

- Gains and losses have to be treated differently. In particular, there are individual reference points (zero points) that divide losses and gains: If we can expect a $60,000 salary, a salary of $40,000 is considered a "loss." The reference point dividing losses and gains might lie somewhere in between (e.g., at $55,000).
- Human behavior toward gains is risk aversive. To give an example: Having choice G, most people (but not all) would prefer the sure gain (alternative G1).

(G) G1 $50 for certain (a gift) or
G2 a gamble with a 50% chance of getting $100 or nothing.

- Human behavior toward losses is risk seeking. To give an example: Having choice L, many people (but not all) would refrain from the sure loss (alternative L1) and choose the more "risky" alternative L2.

(L) L1 a fine of $50 for certain
L2 a gamble with a 50% chance of a $100 fine or of paying nothing.

Figure 5.1 illustrates the assumptions in the model of Kahneman and Tversky. Today, prospect theory is the most powerful extension or revision of the expected utility approach (see Yates, 1990).

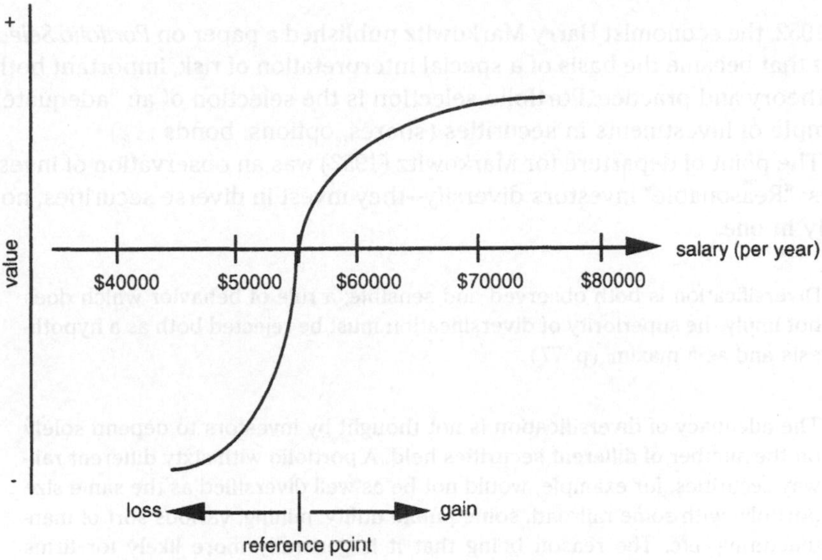

FIG. 5.1. Human decision making under risk (according to Kahneman & Tversky's [1979] prospect theory). This figure shows human decision making under risk, here in the case of decisions for jobs with different salaries (other job conditions being equal). According to Kahneman and Tversky's prospect theory (1979), we have to distinguish between gains and losses. The division between gains and losses is defined by a reference point that depends on the individual and the decision that is to be made. In our example, accepting a job with a salary of less than $55,000 a year is considered a loss (other conditions being equal). Human behavior toward possible gains is "risk aversive": A job with a salary reasonably over $55,000 a year would be preferred to the uncertain chance of a much better paid job. Human behavior toward possible losses tends to be "risk seeking": The uncertain chance for a better job would be preferred to a job with a salary reasonably below $55,000 a year. The difference in risk behavior is shown by the two curves; the vertical axis indicates the psychological value of a job salary (the value replaces the utility of the expected utility approach). The curve for gains is not as extreme as the one for losses and indicates satisficing (in our example, we are content with a job reasonably over $55,000; we won't strive for a top-paid job). Kahneman and Tversky also proposed a new measure for probabilities ("decision weight"), taking into account human peculiarities in understanding probability (e.g., overestimating the very low probabilities). *Note.* From "Prospect Theory: An Analysis of Decision Under Risk" by D. Kahneman and A. Tversky, 1979, *Econometrica, 47.* Copyright 1979 by Econometric Society. Reprinted by permission.

Markowitz: Risk as Variance

In 1952, the economist Harry Markowitz published a paper on *Portfolio Selection* that became the basis of a special interpretation of risk, important both in theory and practice. Portfolio selection is the selection of an "adequate" sample of investments in securities (shares, options, bonds . . .).

The point of departure for Markowitz (1952) was an observation of investors: "Reasonable" investors diversify—they invest in diverse securities, not only in one.

> Diversification is both observed and sensible; a rule of behavior which does not imply the superiority of diversification must be rejected both as a hypothesis and as a maxim. (p. 77)

> The adequacy of diversification is not thought by investors to depend solely on the number of different securities held. A portfolio with sixty different railway securities, for example, would not be as well diversified as the same size portfolio with some railroad, some public utility, mining, various sort of manufacturing, etc. The reason being that it is generally more likely for firms within the same sector to do poorly at the same time rather than for firms in different industries. (p. 89)

Markowitz (1991) concluded that the expected utility approach cannot explain diversification—it can only advise on investing in the most promising:

> To maximize the expected value of a portfolio, one needs only to invest in one security—the security with the maximum expected return (or one such, if several tie for maximum). Thus action based on expected return only (like action based on the certainty of future) must be rejected as descriptive of actual or rational investment behavior. (p. 470)

From the point of view of the expected utility approach, it may be advisable to keep on trying (e.g., to buy many lots until one lot wins) or to avoid severe losses (e.g., by investing some money in risky assets, some other money in risk-free assets), but not to diversify as an optimum, efficient strategy. Markowitz took investors seriously—perhaps more seriously than they did themselves. For him, the problem of portfolio selection was how to invest reasonably and not how to speculate effectively. In his 1990 Nobel lecture, Markowitz (1991) emphasized: "Portfolio theory considers how an optimizing investor should behave" (p. 469).

Markowitz proposed a mean-variance model. The mean in his model is the expected return (of a portfolio) as we find it in the expected-utility model. Variance is a statistical characteristic for the amount of diversity in a sample: For example, when we produce nuts of one kind, they can differ slightly in their actual diameters. Thus, the diameter shows variance—that is, divergence from a mean diameter. However, we have to keep the variance small because the nuts have to fit the screws, which might differ slightly in size. Thus, the variance of a portfolio is a measure of the divergence of the possible returns. Markowitz conceived of risk as variance.

We can explain the variance concept in relation to shares. A share is a capital investment in an enterprise. There are several forms of returns from a share: (a) the dividend, and (b) the price gain. Usually the dividend is paid periodically to the shareholders. The dividend depends on the yields of the enterprise in that particular period. As the yield varies, the dividend varies, too. A share with high variances in the dividend is not preferable because of the uncertainty of returns (see Fig. 5.2). Similar assumptions hold for the price of a share and its price gains. Yet even a share with high risk (variance) might be preferable when we can expect high returns (means). As Markowitz (1952) put it: "The investor does (or should) consider expected return a desirable thing *and* variance of returns an undesirable thing" (p. 77). Markowitz's concept of risk as variance is based, on one hand, on the observable necessity for diversification of risks and, on the other hand, on reasonable, subjective beliefs. Markowitz (1952) insisted that portfolio selection starts with relevant beliefs about the future performances of available securities:

> The process of selecting a portfolio may be divided into two stages. The first stage starts with observation and experience and ends with beliefs about the future performances of available securities. The second stage starts with the

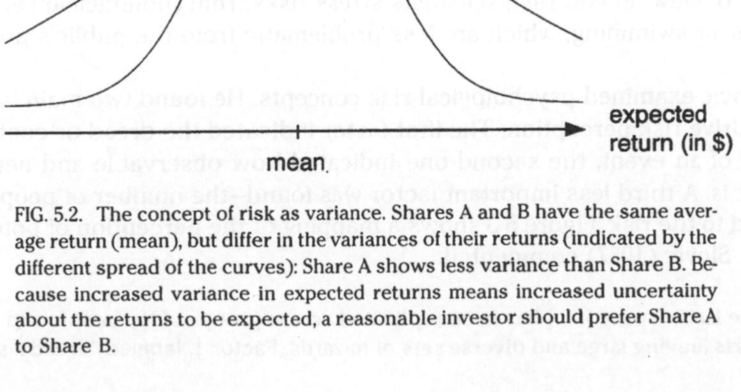

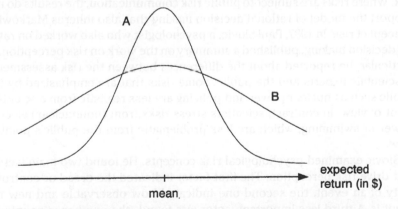

FIG. 5.2. The concept of risk as variance. Shares A and B have the same average return (mean), but differ in the variances of their returns (indicated by the different spread of the curves): Share A shows less variance than Share B. Because increased variance in expected returns means increased uncertainty about the returns to be expected, a reasonable investor should prefer Share A to Share B.

relevant beliefs about future performances and ends with the choice of portfolio. (p. 77)

In his Nobel Prize lecture almost 40 years later, Markowitz (1991) stressed this point again:

> In discussing uncertainty [. . .], I will speak as if investors faced known probability distributions. Of course, none of us know probability distributions of security returns. But, I was convinced by Leonard J. Savage, one of my great teachers at the University of Chicago, that a rational agent acting under uncertainty would act according to "probability beliefs" where no objective probabilities are known; and these probability beliefs or "subjective probabilities" combine exactly as do objective probabilities. This assumed, it is not clear and not relevant whether the probabilities, expected values, etc., I speak of [. . .] are for subjective or objective distributions. (p. 470)

Thus, the mean-variance model works as if all investors had perfect information about the market (once again, for Markowitz, investing is not gambling). This is one reason that the mean-variance model is often thought to imply an equilibrium between investments and investors. Markowitz is only concerned with the reasonable investor—who is no gambler, but tries to optimize and secure investments.

Some Basics of the Psychology of Risk Perception

Psychologists investigated the common perception and understanding of risk. Where risks are subject to public risk communication, the results do not support the model of rational decision making that also inheres Markowitz' concept of risk. In 1987, Paul Slovic, a psychologist who also worked on rational decision making, published a summary on the work on risk perception. In particular, he reported about the differences between the risk assessments of scientific experts and the public. Some risks that are emphasized by the public such as nuclear power and policing are less realistic from a scientific point of view. In contrast, scientists stress risks from (nonnuclear) electric power or swimming, which are less problematic from the public's point of view.

Slovic examined psychological risk concepts. He found two main factors that drive risk perception. The first factor indicated the dread or controllability of an event, the second one indicated how observable and new the event is. A third less important factor was found—the number of people exposed to the risk. Figure 5.3 shows a mapping of the perception of potential risks. Slovic (1987) commented:

> The factor space [. . .] has been replicated across groups of lay people and experts judging large and diverse sets of hazards. Factor 1, labeled "dread risk,"

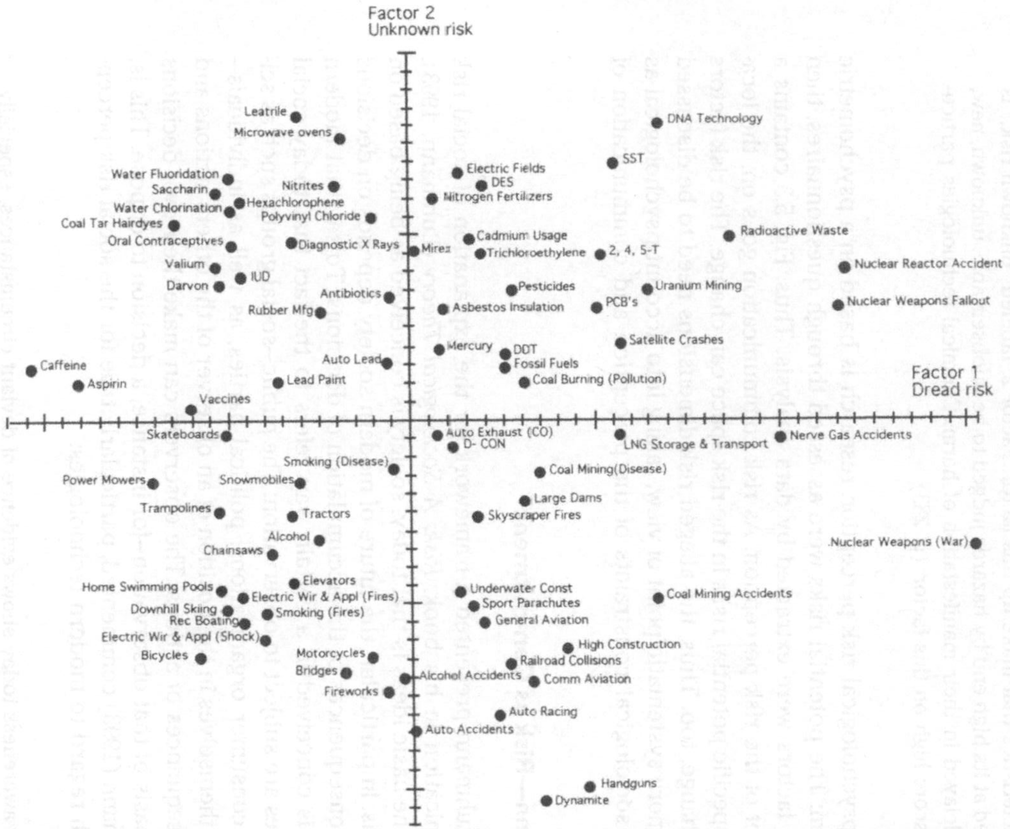

FIG. 5.3. Risk perception. Location of some hazards on Factor 1 (uncontrollable, dread risk) and Factor 2 (unobservable, unknown risk) according to risk perception research (Slovic, 1987). The factors were derived from the relationships among 18 risk characteristics. The attitude toward laypersons of regulation of a potential risk is strongly correlated with the dread (controllability) factor. *Note.* From "Perception of Risk" by P. Slovic, 1987, *Science, 236*, pp. 282–283. Copyright 1987 by American Association for the Advancement of Science. Adapted by permission.

is defined at its high (right-hand) end by perceived lack of control, dread, catastrophic potential, fatal consequences, and the inequitable distribution of risks and benefits. Nuclear weapons and nuclear power score highest on the characteristics that make up this factor. Factor 2, labeled "unknown risk," is defined at its high end by hazards judged to be unobservable, unknown, new, and delayed in their manifestation of harm. Chemical technologies particularly score high on this factor. (p. 283)

The psychological risk perception research is based on a psychometric paradigm: The potential risks were assessed through questionnaires, then the risk factors were extracted by data analysis. Thus, Fig. 5.3 contains a snapshot of the risk perception. As risk communication goes on, the location of specific potential risks in the risk space can change. The risk factors might change, too. Thus, the alleged risk dimensions need to be discussed from a more systematic point of view, taking into account psychological as well as sociological constraints of the perception and communication of risks.

Luhmann—Risk as Commitment

Niklas Luhmann presented a framework for the explanation of social risk communication in his book, *Risk: A Sociological Theory* (Luhmann, 1993). One of the basic ideas is that today, society is conceived as being based on decisions. In particular, the future of modern society depends on decisions or the consequences of the accumulation of decisions. To say that modern society is conceived in a certain way refers to the fact that today social processes are subject to observation: The public–social groups such as scientists, consumer organizations, political parties, as well as individuals—can put themselves in the position of an observer of the others' actions and the consequences of actions. The observers can make their own decisions on the basis of that observation–for instance, a decision to oppose. This is, as Luhmann (1993) commented, particularly true for the social risk perception with regard to modern technologies:

> [R]isk awareness today shows evidence of deviant circumstances, especially a fascination with the possibility of extremely improbable occurrences, which–when they do happen–constitute a disaster. This cannot be explained alone by the fact that technology offers such possibilities [. . .]. The explanation is likely to be that nowadays people or organizations–that is to say decisions–can be identified as the root cause. It makes sense to oppose. (p. IX)

For example, the environmental situation in the neighborhood of an industrial site or a nuclear power station is subject to observation. Every change or degradation is likely to be seen in connection with technical malfunctions. Therefore, it cannot be surprising that the research on risk perception not only revealed the risk factor of controllability but also one of observability, as shown in Fig. 5.3. Observation follows the rules of attribution (discussed in chap. 3). Luhmann recalled that, "actions and consequences of actions are a problem of attribution and are made perceptible by attribution" (p. 67). In the case of social risk communication, attribution couples "two temporal contingencies": an event and a loss (p. 17). In this way, possible future events are conceived as being dependent "on decisions to be made at present." Luhmann explained:

> For we can speak of risk only if we can identify a decision without which the loss could not have occurred. It is not imperative for the concept [of risk in our time, HAM] [. . .] whether the decision maker perceives the risk as a consequence of his decision or whether it is others who attribute it to him; and it is also irrelevant at what point in time this occurs—whether at the time when the decision is made or only later, only when the loss has actually occurred. (p. 16)

One central argument underlying Luhmann's explanation of risk starts with the observation that today the concept of risk has gained a universal, critical power in social communication, as seen in

> [t]he question of who or what decides whether (and in which material and temporal contexts) a risk is to be taken into account *or not*. The already familiar discussions on risk calculation, risk perception, risk assessment and risk acceptance are now joined by the issue of selecting the risks to be considered or ignored. (p. 4)

The argument goes: The concept of risk would not have gained such a universal, critical power if it were not functional to today's understanding of decision making. The concept of risk, as has already been said, can couple events and losses. It does so by distinguishing between events that are consequences of decisions and events that are not. Therefore, Luhmann distinguished between risk and danger; risks refer to decisions, danger does not:

> The distinction [between risk and danger] presupposes [. . .] that uncertainty exists in relation to future loss. There are then two possibilities. The potential loss is either regarded as a consequence of the decision, that is to say, it is attributed to the decision. We then speak of risk—to be more exact of the risk of decision. Or the possible loss is considered to have been caused externally, that is to say, it is attributed to the environment. In this case we speak of danger. (p. 22)

Due to the modern view, events are socially attributed as consequences to decisions (of persons, institutions, cultural habits, etc.). Hence, "modern society considers danger from the point of view of risk and takes it seriously only as risk" (p. 27). Earthquakes seem to be inevitable dangers, but: "Even if it is only a question of danger in the sense of natural disaster, the omission of prevention becomes a risk" (p. 31). Luhmann concluded: "modern society represents the future as risk" (p. 37). Also Slovic, in his summary on risk perception as represented in Fig. 5.3, did not mention a single danger in the sense of Luhmann: All risks Slovic reported about stem from human or social activities, such as police work, electric power, or vaccinations.

According to Luhmann, the distinction between risk and safety has become obsolete, in particular in risk communication on new technologies. Even safety experts today state that there is no absolute safety and some—but improbable—risk always remains. Science as well as models of rational decision making help determine the reasonable potential risks, which then can become objects of social risk communication—that is, according to Luhmann, a public dispute on society's future. In Luhmann's words:

> If there are no guaranteed risk-free decisions, one must abandon the hope that more research and more knowledge will permit a shift from risk to security. Practical experience tends to teach us the opposite: the more we know, the better we know what we do not know, and the more elaborate our risk awareness becomes. The more rationally we calculate and the more complex the calculations become, the more aspects come into view involving uncertainty about the future and thus risk. (p. 28)

Synopsis

At this point, we are not in a position to present any kind of integration of the two concepts of risk. Instead we point to different assumptions or conditions that underlie the two approaches.

(a) The *as-if-we-all-knew clause*. This is the assumption that everybody has access to the same information or, at least, that we have sufficient information to make a decision. Portfolio selection theory as well as large parts of the theory of rational decision making cannot function without that assumption. Decision trees on which therapy to choose (e.g., Fig. 2.8) or which stock to select would be senseless without assuming that the alternatives and probabilities do not lack any reality—also when we speak of expected returns or expected utilities that have to be realized in an uncertain future. The risk concept by Luhmann does not make use of the *as-if-we-knew* clause. According to Luhmann, future has to be decided on, however sufficient our knowledge for decision making or rational the decisions might be.

(b) The *time-&-commitment clause*. From the point of view of risk communication (especially for Luhmann), decisions are accounted to somebody—it may be a person or an institution. That implies that deciding means a commitment to factual decisions. Usually decisions face narrow time constraints: There is often no time to analyze every alternative—and often it is unnecessary. In the remainder of the book, we address this concept of risk as risk as commitment. In contrast, portfolio selection theory as well as the theory of rational decision making can function without the time-&-commitment clause; there is no need for decisions to be made.

The concepts of risk by Markowitz and Luhmann do not refer to the same subject. Only Markowitz is directly concerned with financial markets, whereas Luhmann's concept mainly refers to political and organizational decision making—but does not exclude other fields. If we try to view financial markets (without having them adequately introduced yet) from the perspective of each of the two concepts, we see two different "worlds." From the Markowitz perspective, financial markets consist of relatively independent clusters of investment chances (securities). From the Luhmann perspective, financial markets are networks of actions by actors who are dependent by mutual observation and compete for control (of the consequences of their decisions). In the Markowitz world, we can live with markets being driven by random processes, whereas in the Luhmann world, we try to rule out randomness as far as possible. Thus, dealing with risk according to Markowitz or Luhmann, respectively, results in completely different dynamics and leads to different standards. Markowitz clearly advised us to diversify risks (investments) and never stick to one single option. Luhmann did not provide advice, but would confirm the relentless tendency of market participants toward the gain of control of the market's processes. In a nutshell, Markowitz said: Risks are defined by the markets, whereas Luhmann would say: By taking your chance, you can contribute to the definition of the market risks.

In summary, we have seen two different concepts of risk: risk as variance and risk as commitment. On the one hand, we find a calculation of risk based on means and variances, as if there was perfect knowledge. The reasonable investors follow reasonable subjective beliefs, Markowitz insisted (*as-if-we-knew* clause). On the other hand, we find some kind of commitment to a decision (*time-&-commitment* clause). From a psychological point of view, people fear risks because they perceive them as having uncontrollable or unobservable consequences. From a sociological point of view, there is no risk but through attributable social decisions. Modern society represents the future as risk, according to Luhmann. If we view financial markets in the light of these two concepts of risks, different recommendations follow: On the one hand the advise to diversify (Markowitz), on the other hand the need for controlling the consequences of one's own decisions (Luhmann).

5.2 CAN THERE BE EXPERTISE IN FORECASTING FINANCIAL MARKETS?

The worldwide crash of stock markets in 1997 and 1998 were not unprecedented. About 10 years before, on October 19, 1987, the U.S. stock market almost collapsed. The Dow Jones Industrial Average fell 508 points to 1738.74, a 22.6% decline in a single day. That exceeded the 12.8% fall on the so-called Black Friday on October 25, 1929, when a worldwide economic crisis started. After the 1987 crash, economist Robert J. Shiller (1990) sent a questionnaire to institutional and individual investors, asking for their "own personal experience" during the crash:

> Investors were asked: "Did you think at any point on October 19, 1987, that you had a pretty good idea when a rebound was to occur?" In the United States, about a third of both individual and institutional investors said "yes," far more than the proportion who traded on October 19. The questionnaire then asked: "If yes, what made you think that you knew when a rebound would occur?" Again, respondents were given a space to write in their own answers. What struck me in reading the answers of US respondents was the frequency with which people wrote "intuition," "gut feeling" or the like. (p. 57)

We can assume this: If one third of the traders really had foreseen the "rebound" (by virtue of "intuition" or not), they would have taken preventive steps and the stock market would have moved earlier or in another direction.

Capabilities in forecasting financial markets would be highly rewarded. Discussing the "narrative of economic expertise," McCloskey (1990) pointed out that forecasting economic processes is "magical." Moreover, as McCloskey claimed, economics is generally *not* concerned with forecasting. Economic processes are "human events," meaning they result from interrelated human action: "But the forecasting of human events—which is *not* the main activity of economics—has always been magical" (p. 97).

The main activity of economics is modeling—for instance, modeling the interrelationship between demand and supply. However, as philosopher Goodman (1954) remarked, relationships in models can be easily interpreted as causal assertions. Causal assertions qualify for prediction—for instance, about future investment rates or inflation. Nevertheless, McCloskey's (1990) verdict of economic forecasting is strict: "[F]orecasting the future and manipulating it are identically magically" (p. 98).

McCloskey's verdict of economic forecasts seems too strict. On the one hand, we do not really know whether the simple fact that economic processes are connected to human action make them unpredictable. If we base any conclusions on this assumption (as in McCloskey's verdict), we have to make use of an *as-if-we-all-knew* clause. On the other hand, if we accept that economic decision making is based on incalculable uncertainty, then we can also render McCloskey's verdict as productive: Let us connect forecasting and decision making in a dynamic manner–this is what we call *planning*. In this chapter, we see the close relationship between forecasts and planning in financial management.

Let us turn to the question of: if and how expertise in forecasting financial markets is possible at all. Before being able to discuss individual expertise in forecasting financial markets, we have to introduce the main characteristics of this domain,

- first and above all: money;
- second, financial markets;
- third, theory and research on market participation that reveal some characteristics of the behavior of financial experts.

Money

The domain of finance is essentially determined by some of the characteristics of money. Money in general is a medium of exchange. Commodities (books, cars, etc.) and services (housekeeping, surgery, etc.) as well as rights (copyrights, hunting permits, etc.) or duties (debts, support of family members, etc.) can be exchanged for money. According to economist Auerbach (1982), the use of money–as a medium of exchange–depends on the uncertainty of future exchanges. If we had perfect knowledge about all our future exchanges, money would become useless; According to Auerbach:

> The nature of the special kind of uncertainty can be understood by [. . .] imagining the opposite environment–complete certainty about all future receipts and expenditures. Imagine that you know for certain that you would buy a bunch of turnips for $3.26 12 years from now on July 14, after receiving your $6379.56 paycheck on July 12. Your knowledge extends in this way through every last penny of receipts and expenditures to the moment of your death and even beyond [. . .]. Ask yourself, "if everyone had this knowledge, would they need to carry checkbooks, currency and coins?" All future transactions could be recorded in a giant computer bank, and changes in bank balances arising from here to the knowledge horizon could be cleared today. (1985, p. 10)

Put another way, the importance of the exchange function of money is based on our freedom to engage in an exchange or not (to buy a book or to borrow it).

The concept of money has–at least–two different aspects: money as value and money as debt. These aspects do not necessarily exclude one another however–in their extreme versions–they cause different attitudes toward money. Thinking of money, the aspect of value might be more salient. The concept of money as value means that there is some basic value inherent in the money, such as in silver or gold coins, or behind the money that covers it, such as the gold standard in the late 19th century or today's GNPs. From that point of view, it is reasonable to store money, especially for future transactions. However, the concept of money as debt means there is some kind of a contract in which one person or institute promises (sometimes by a signatured bill) to return later what he or she has gained now. Debt economy creates credits and debits that can be accounted for without any exchange of physical money. Checks, bills of exchange, as well as paper money show this characteristic of a debt; in the words of Weber (1947): "From a legal point of view, paper money may consist in an officially redeemable certificate of indebtedness" (p. 291). From that point of view, it is advisable to reclaim or convert (debt) money as soon as possible, one reason being that its value depends on the trustworthiness of the debtor: For example, the issuer of the bill of exchange may turn out to be a cheat. We find examples for both aspects of money in primitive cultures (see e.g., Einzig, 1966), although the value aspect—embodied in coconuts, beads, pigs, cattle, and so on—by far prevails.

From a psychological point of view, money has motivational and cognitive aspects. Generally, as to the motivational aspect, money is a so-called *secondary reinforcer*. A reinforcer is something (an event, an object) that reinforces certain behavior of persons or animals: We can teach dolphins tricks by reinforcing them through successively feeding fish each time their behavior is closer to the trick we aim at (e.g., jumping through a hoop). Similarly, laughter reinforces a person's joke telling. In our examples, fish feeding and laughter are reinforcers. Primary reinforcers meet basic needs such as hunger or affiliation. In contrast, secondary reinforcers are learned; they substitute primary reinforcers in that they reinforce behavior even in absence of a basic need.

Money as a secondary reinforcer makes us render services (e.g., as employees) or hand out goods (e.g., a valuable watch) without direct satisfaction of our basic needs. As a secondary reinforcer, money can reinforce its own accumulation—that would be impossible for primary reinforcers.

Second, as to the cognitive aspect, money is abstract. We can measure almost everything in terms of money—from general goods (foods) to personal relations (marriages). Coins and paper money may still have a concrete substance like metal and paper, but bank account figures do not. If some of your paper money burns, usually the money is lost. Yet if your bank burns, your money will not be lost. Bank accounts—as such—neither exist on paper nor in computers because the form of representation can change without damage to the account. Your bank account exists on the agreement that the bank will repay its debt to you, there are general conventions, a traceable history of the account, and, of course, documents such as account opening forms and statements. Thus, by using money we

TABLE 5.1
Willingness to Pay for Car Safety Features
That Prevent Injuries Resulting in Death (Means)

	Starting Point £ 25	*Starting Point £ 75*
Minimum	£ 87	£ 232
Best	£ 113	£ 265
Maximum	£ 149	£ 350

Note. The interviewer also asked for a single amount between the maximum and minimum at which the responders would find it hardest to decide whether to pay; this amount is shown as "best" estimate. Data from Jones-Lee and Loomes (2000) in *Conflict and Tradeoffs in Decision Making* by E. U. Weber, J. Baron, & G. Loomes, 2000, New York: Cambridge University Press. Copyright 2000 by Cambridge University Press. Adapted by permission.

can reduce the variety of goods into an abstract comparable entity—an accounting unit. The abstract character of money that allows for general comparability also renders the value of money totally dependent on any psychologically relevant reference points. The effect is also known as anchoring bias (e.g., Tversky & Kahneman, 1982). For example, British psychologist Loomes (1997) reported on his experiments about the willingness to pay for car safety features that prevent injuries resulting in death. As Table 5.1 shows, if the question started with £25, responders were at the utmost maximum willing to pay £149. Yet this maximum was far below the minimum of £232 responders were willing to pay if the interviewer started the question with £75.

Financial Markets

We now turn to financial markets. Money markets are financial markets where short-term debts (less than 1 year) are traded. Capital markets are financial markets where long-term securities such as stocks and bonds are traded. Usually we also have to distinguish between primary markets where new financial assets are sold and secondary markets where these assets are resold. We focus on secondary markets, in particular on stock exchanges, but we also take a look into the option and future markets and the international currency exchange market. These financial markets develop steadily. To start with, we describe three general trends.

1. *Globalization.* Computer and information technology have linked international financial markets closer than ever. The worldwide crash of the stock markets in October 1997 was a direct consequence of the financial crisis in Asia and the collapse of national Asian stock markets, in particular that of Thailand. The interlinkage is complete in currency markets. During the business week, the world's important currencies (US$, Yen, EURO, UK£, SFr.) are traded almost around the clock on one of the three most important cur-

rency markets. When Tokyo closes, London is about to trade; when London closes, New York is still busy.

2. *Securization.* Securization means the process by which traditional bank loans have been replaced by tradeable securities. A security usually refers to a piece of paper that serves as evidence for property rights and that may be transferred with all rights and conditions to another investor. Stocks and bonds are such securities. Instead of lending capital from a commercial bank, a company may choose to issue new stocks or corporate bonds. Securities are more liquid, because they can be traded. Securization intensifies market activity.

3. *Competition of the financial markets (market places) for new financial instruments.* As a byproduct of ongoing globalization, investors can now choose not only the kind of investment, but also the markets where they want to invest. Internationally operating corporations can also choose where to trade their shares. Thus, not only the participants but also the market places compete. New financial instruments are created, and markets can specialize on that specific instrument, derivates and market-index funds being the most prominent ones.

Differentiation and interdependence of the financial markets presumably increase the complexity of the whole system of financial markets. In the aftermath of the 1987 stock market crisis, an international expert meeting at ministerial level tried to identify systemic risks caused by the interdependence of the international financial markets (OECD, 1991). Because neither the concrete risk nor the real complexity are known, we prefer to speak of *systemic uncertainty*—that is, incomplete knowledge about the causal connections in the financial market system. One example for systemic uncertainty is the effect of arbitrage with index futures:

> Index arbitrage exploits discrepancies between the price of an index future and the cash price of the component stocks in the index. Thus, if at a particular time—allowing for transaction and carrying costs—the future price is cheap, arbitrageurs will buy futures and sell the corresponding cash stock, thereby securing a risk-free return.

As to arbitrage with index futures, the 1991 OECD report contains two totally diverging assessments:

(a) Index arbitrage supports the balancing of different financial markets; the absence of such arbitrage was one cause of the 1987 collapse (the London point of view);

(b) On the contrary, index arbitrage can be a source of market destabilization—by transmitting volatility from the derivate market to the cash market (the New York point of view).

Theory and Research on Market Participation

In theory, financial markets can be characterized by their efficiency (see Fama, 1970, 1991). "A (perfectly) efficient market is one in which every security's price equals its investment value at all times" (Sharpe et al., 1995, p. 105). There are several forms of market efficiency:

(a) In case of *weak-form efficiency*, security prices reflect all previous information on prices. Thus, no investor can consistently earn excess returns with trading rules based on historical price or information.
(b) In case of *semistrong-form efficiency*, security prices reflect all public information. Thus, no investor can consistently earn excess returns by using information that is publicly available.
(c) In case of *strong-form efficiency*, security prices reflect all relevant information on prices (public and private). Thus, no investor can consistently earn excess returns by using information that is available.

Financial markets are not efficient in the strong sense that any relevant information will be instantaneously reflected by security prices. Thus, investors are still able to earn excess returns from using relevant information (about firms or interest rates) not yet reflected by the market. This information, of course, only allows for short-term predictions. Financial markets seem to be weak-form efficient, excluding excessive gains from the use of historical information (Sharpe et al., 1995).

A basic idea related to market efficiency is that of an equilibrium between risk and return of investments—an idea theorized in the Capital Asset Pricing Model (CAPM; Sharpe 1964).[1] In short, the risk return relationship can be roughly described as: The greater the risk, the greater the expected

[1]The central term of CAPM is the beta coefficient that represents the covariance of a security with a market and measures risk (cf. Sharpe et al., 1995, chap. 10). The beta coefficient β_{iM} for an investment i and a market M can be expressed by:

$$\bar{r}_i \quad r_f \quad (\bar{r}_M \quad r_f)\ \beta_{iM}$$

$$\beta_{iM} \quad \frac{\sigma_{iM}}{\sigma_M^2}$$

$\bar{r}_i$: expected return for investment i
r_f: riskfree return (e.g., securities from the US Treasury)
$\bar{r}_M$: expected return of the market portfolio (its variance σ_M^2)
σ_{iM}: covariance of the investment and the market portfolio

Despite all limitation, the beta coefficient is in wide use as a quick risk estimation for a security (especially shares). As to options and other derivates, the standard risk measure in use today (instead of the CAPM β) is given by the Black–Scholes (1973) formula.

return. The reason is: No reasonable investor would accept risk without expected return. According to the theory, the stock market is an efficient investment where the volatile (i.e., risky) stocks correlate with expected high returns. That is the theory.

In reality, stock markets have not only proved to be not strong-form efficient, but also show some anomalies, one being the evidence for market overreaction. As early as 1936, Keynes stated: "Day-to-day fluctuations in the profits of existing investments, which are obviously of an ephemeral and nonsignificant character, tend to have an altogether excessive, and even absurd, influence on the market" (1964, pp. 153–154; cited after DeBondt & Thaler, 1985, pp. 793–794).

It is often claimed that stock price indexes seem to vary more than we might expect from a rational point of view. The changes cannot be attributed to new information (Shiller, 1981). Investors seem to attach disproportionate importance to short-term economic developments. In the long run, the market regulates stock prices: Extreme movements in stock prices will be followed by subsequent price movements in the opposite direction. Haugen (1995), a critic of the theory of market efficiency, commented:

> Investors mistakenly tend to project a continuation of abnormal profit levels for prolonged periods in the future. The stock prices for successful firms tend to become overvalued. Unsuccessful firms tend to become undervalued. Then, as the process of competitive entry and exit drives corporate performance to the mean faster than anticipated, the overreactions are corrected. The stock of the once profitable firms produce low returns. The converse is true for their unprofitable counterparts. (p. 24)

Who overreacts, the market or the market participants? The market is an aggregation of the behavior of the individual participants, without the traders the market would not exist. However, there are phenomena on the market level (such as business cycles) that cannot be explained by reducing it to the behavior on the individual level. As to stock markets, there is the saying that no one can "beat the market" in the long run. That means: The most profitable investment would be the market itself. As we see, expert selected investments hardly do better than the market (as measured by indexes like the Dow Jones Industrial Average). Thus, market definitions do not need reference to individual behavior: "A market, according to the masters, is the area within which the price of commodity tends to uniformity, allowance being made for transportation costs" (Stigler, 1942/1966, p. 77).

We do not want to question or blur the theoretical and practical distinction between markets and individuals. We only want to point to the general assumption that is implicit in the research on overreaction and many other anomalies: the assumption of a best, rational behavior. Overreaction means a reaction that is too excessive compared to rational standards or theory-based predictions. Thus, DeBondt and Thaler (1985, 1987), who investigated overreaction in stock markets, argued that the investor's behavior is not rational. As discussed in the framework of cognitive economics (chap. 2, this volume), we leave the point of rationality open to discussion. It is impossible to generally decide whether the individual's behavior is not rational or the researcher's model of rationality is misleading.

We now turn to the functions of the market participants. We will not try to give an overview of the investors and market players because there are too many of them, and the various types depend on the particular markets and their legal frameworks (e.g., brokers are common in the United States and quite uncommon in Germany). Instead, we describe possible functions seen in relationship to the different forms of investment. We can isolate four functions:

1. *Lending and borrowing capital.* A corporation or a bank may be in need of capital and thus sells bonds or shares on the market. Here we find the reasonable investor, as described by Markowitz: an investor who tries to find an optimum portfolio by balancing risks and expected returns. There are, of course, many private investors, often called retail investors. But today, we find more and more institutional investors like pension funds or insurance companies who concentrate financial power in their hands.

2. *Market making (including arbitrage).* Not every security (stocks, bonds, options, etc.) is traded all the time. However, market participants can buy particular securities—on supply—and resell them on demand, thus making a market and ensuring that every time that particular security can be sold or bought without having a complementary counterpart that wants to buy or sell. Market makers are common in trading options. They exploit the spread between the price they bought the security for and the price they sell for. Similarly, arbitrageurs exploit the price differences between different markets: They buy at one market and resell on the next.

3. *Risk diversification (including insurance).* Options and futures can be used as insurance against risks. For instance, in 1997, a Swiss manufacturing firm receives an order for two turbines for a Chinese dam. The turbines are to be delivered in January 1998. The total purchase price is $10 million and is payable on delivery. The deal may be worth about 15 million Swiss francs. Yet if the dollar falls before January 1998, the firm risks losing the profit from that particular deal. In this case, it can offer or buy a future contract to sell $10 million for 15 million Swiss francs in January 1998. Thus, if the dollar falls, the firm can resell the $10 million; the future insures the profit. The insurance of businesses by options or futures is usually called *hedging*. Hedging is a specific case of risk diversification through financial markets. An entrepreneur or investor can sell risks in the form of options or futures, offering a premium for those who take over the risk.

4. *Speculation*. Investments can also be used for speculation. To raise the expected return on investment, one can use leverage effects. For example, instead of investing one's own capital into stocks, one invests in a loan. With the borrowed money, one can buy more shares. The expected return rises, as does the potential loss. The leverage can be further expanded by buying options on a stock rather than the stock itself. If the price of the stock rises, the price of the call options rises much more. Thus, a speculator can win and lose a fortune. For many, speculation might be the motivation to engage in financial markets, thus investing and taking on risk. We can suppose that speculation shapes the financial markets as much as the other functions do.

The system of these financial market functions is, to some extent, recursive. Every regulation or instrument for risk management—hedging, clearing—can be used to take over even riskier investments, for example, engaging in leverage. Indexes (Dow Jones, DAX, Nikkei) that, on one hand, help assess the relative risk of investments become, on the other hand, objects of bets with futures. In this way, huge debts are amassed. As to outstanding option trading positions, the International Monetary Fund (IMF) resumed in its 1995 reports:

> [A]t the end of 1994, approximately $500 billion of replacement value was outstanding, compared with the capital base of less than $200 billion of the 12 largest dealers, who together are responsible for the vast majority of over-the-counter transactions in derivates. (Folkerts-Landau & Ito, 1995, p. 18)

The report of the OECD meeting on systemic risks in the aftermath of the 1987 collapse of the stock markets recommended "improving settlement systems" (for the settlement of financial contracts) and "developing an internationally coordinated supervisory system for internationally active intermediaries in security markets" (OECD, 1991, p. 50). The report also admitted the new potential risk of such regulations: "Official readiness to relax monetary conditions may, however, itself encourage excessive buoyancy in securities markets, and expectations of support for failing firms may encourage imprudent behavior of intermediaries."

Financial Forecasting Expertise

After having displayed financial markets as an environment for decision making, let us now turn to the topic of individual expertise. In 1972, Stael von Holstein published an experiment on the forecasting of stock prices at the Stockholm Stock Exchange. The experiment included five groups of persons with varying degrees of experience of the stock market:

- bankers working in the stocks and bonds department of a Stockholm bank (hereinafter *bankers*);
- a "heterogeneous group of experts," some actively connected to the stock exchange and some doing academic work on it (*stock market experts*);
- people associated with the Institute of Mathematical Statistics of the Stockholm University (*statisticians*);
- teachers of business administration at Stockholm University (*business teachers*); and
- students of business administration (*students*).

The task was to predict the price of 12 stocks over 2 weeks. For every stock, each participant had to state the probability of each of the following five cases: (a) the price decreases by more than 3%, (b) the price decreases by more than 1% but by 3% at most, (c) the price changes by 1% at most, (d) the price increases by more than 1% but by 3% at most, or (e) the price increases by more than 3%. The experiment consisted of 10 sessions; each time the participants predicted the change in price for the 12 stocks over a 2-week period. The stocks were quoted on the Stockholm Stock Exchange and represented various industries, mostly large companies (by Swedish standards).

The results were astonishing. Only 3 of 72 participants reached scores that were better than random. On average, the statisticians ranked first, the experts second, and the bankers worst. The students ranking third were more successful than the business teachers ranking fourth.

> Statisticians > stock market experts > students > business teachers > bankers

Is there no real expertise in predicting stock prices? One might suspect that the experimental task Stael von Holstein used was not realistic enough. Usually the stock price is judged directly ("Stock X will rise in the next Y months to price Z"), rather than in terms of a probability distribution (as in cases b to d). The task needed general statistical skills rather than mere experience, therefore being to the statisticians' advantage. One might also restrict the potential domain of expertise in forecasting stock prices to people actively working at the stock exchange, thus excluding stock specialists within banks. Indeed, four out of the best five individuals in the experiment were stock market experts. Then we would have to ask: What kind of prior experience do stock market participants have that investment bankers do not have?

Since October 1988, the *Wall Street Journal* has been publishing a column called "Investment Dartboard." It appears on a monthly basis. The core idea is to compare the stock selections of experts with ones of darts thrown by employees of the *Wall Street Journal*. Each month, four investment professionals recommend one stock each for the coming 1-month or 6-month period. The column presents the investment professional's picture along with his or her recommendation, the performance of the stocks selected by these experts and by the darts.

Several studies used the "Investment Dartboard" as a database. In a study on the first 43 contests (Sundali & Atkins, 1994), the experts clearly outperformed the dartboard. More interesting, 58% of the stocks recommended by the experts also outperformed the Dow Jones Industrial Average Index. That means that 58% of the stocks recommended by the investment professionals rose by more than the average stock market. Assuming that the random chance for any particular stock price to exceed the performance of the Dow Jones is 50%, the expert judgment adds about 8% in certainty. This increase is statistically significant. In comparison, a medical treatment (e.g., surgery of fractures) with an 8% over random chance of success would be considered a risky treatment rather than a standard one.

Other data are provided by professional advisory services such as Value Line or the Institutional Brokers Estimation System (IBES). A recent study using IBES data (Capstaff, Paudyal, & Rees, 1995) confirms former findings on professional investment forecasting. The study includes forecasts on companies' future earnings per share that have shown to be an important measure also for stock price changes (p. 67). The study shows:

- Analyst's forecasts are quite accurate in comparison to no-change predictions, especially in short-time predictions (less than half a year). Their superiority diminishes over longer time horizons (1 year and longer). Moreover, analysts' forecasts are more accurate in cases where a company's earnings increase.
- Analysts' forecasts show an optimistic bias; 55.4% of the forecasts were higher than the actual value (42.4% lower, the remainder was exact). In 61% of all revisions, analysts had to revise downward.

Financial markets change fast, perhaps too fast for individuals to adjust. From a psychological point of view, there are at least two possible forms of expertise in forecasting financial markets:

(a) *strong-form expertise*: There is a reasonable forecast capability that is based on individual experience and develops according to the general 10-year rule (it takes about 10 years of training and experience to develop expertise).

TABLE 5.2
Accuracy of Expert Earnings Per Share Forecasts,
Based on 52049 IBES Forecasts in UK

Forecast Horizon	*Analysts' Mean Forecast Error*	*Analysts' Forecast Error-Earnings Increased*	*Analysts' Forecast Error-Earnings Decreased*
0 month	11.5%	7.2%	19.6%
5 months	18.4%	9.4%	34.5%
10 months	18.5%	12.4%	41.9%
20 months	23.5%	13.7%	57.5%

Note. Data from "The Accuracy and Rationality of Earnings Forecasts by UK Analysts" by J. Capstaff, K. Paudyal, and W. Rees, 1995, *Journal of Business Finance & Accounting, 22*, pp. 74–75. Copyright 1995 by Blackwell. Reprinted by permission.

(b) *weak-form expertise*: Relative forecast capabilities result from the use of specific, concrete information (e.g., insider information within firms) or specific technology (e.g., computers).

As to financial market forecasts, data do not provide evidence for expertise in a strong sense. There seems to be no fundamental development in the individual expertise for forecasting financial markets. Many who know the "rules" can "play" the financial market. Amateur investment clubs can outperform professional investment firms (Maturi, 1995). This is not so surprising if we take into account that stock markets are not truly efficient either.

We should briefly consider the nature of intuition in predicting financial markets. Professionals as well as researchers might claim that there is a strong intuitive component in predicting financial markets. Top trader Michael Marcus commented: "Gut feel is very important. I don't know of any great professional that doesn't have it. Being a successful trader also takes courage: the courage to try, the courage to fail, the courage to succeed, and the courage to keep on going when the going gets tough" (Schwager, 1989, p. 44). From a psychological point of view, we can suspect that *gut feeling* indicates a specific cognitive strategy in problem solving.

There is much empirical evidence against the validity of gut feeling. For instance, Oskamp (1965) showed that professional self-confidence in judgments increases when more information is available, but without any corresponding increase in judgmental accuracy. Nisbett and DeCamp Wilson (1977) showed, more generally, experts being mostly unaware of what they know. Gut feeling seems to be a case of the illusion of control (see chap. 3, this volume). However, gut feeling does seem to indicate a motivational component of risk taking (courage). It is no wonder that, as Jack D. Schwager (1989) reported in his book *Market Wizards*, almost all top traders started their careers with huge losses.

Where does forecast accuracy come from? Relative forecast capabilities may result in particular from information within firms. In this case, forecasting is confounded with planning. For instance, if a company's management—on the basis of market research and successful first experiences—plans to invest in South Africa, this decision strongly affects future earnings and therefore the stock price. However, if the accounting department subsequently discovers that the company's equity and capital structure do not allow for fast growth, this piece of information can change expected earnings, too. From this point of view, the difference between the reasonable forecast capability for increased earnings and the relative inability to predict decreasing earnings (as in Table 5.2) can be attributed to planning information. Where a company develops according to plan, earnings increase. Otherwise earnings unexpectedly decrease. The changes are reflected by the stock market. Planning information or insider information, respectively, have always been a means of forecasting stock markets. Daniel Drew, a speculator in railroad stocks of the 19th century, is quoted as saying: "Anybody who plays the stock market not as an insider is like a man buying cows in the moonlight" (cited in Wood, 1988, p. 18).

Unintended Effects of Theory and Expert Advice: Volatility

As we have seen, expert information on the expected earnings can be an important signal for investments. As we see, from a social psychological point of view, experts may also serve as mediators for the high volatility of financial markets.

In several experimental studies, social psychologist Andreassen investigated the impact of information provided by the financial press on the behavior of investors and explained how volatility might be produced. First, Andreassen showed that without information, investors tend to regressive trading behavior: They only sell when the stock price rises and only buy after it falls. Large price changes cause increased trading behavior—that is, increased volatility. "Trading on the stock market increases when there are large changes in price levels, and falls when these changes are small" (Andreassen, 1988, p. 371). Then, Andreassen documented that the financial press, such as the *Wall Street Journal*, explains recent changes with good news after price rises and bad news after price falls. In an experimental study, Andreassen demonstrated that such news leads to less regressive trading behavior. The explanations provided by the press are causal attributions, the investor is given a reason (ex post) that the change had to come: For instance, the rising price of a stock might be explained by changes in the management of the corresponding company. Thus, the investor might be more inclined to expect a further rise than in the case of no explanatory information: "[O]ne group [the investors] acts on the attributions provided by another [the press]" (Andreassen, 1987, p. 490).

In an additional experiment, Andreassen (1990) showed that it is information on price changes and not any kind of news that accounts for the explanatory power. Thus, we can conclude: From a social psychological point of view, information explaining changes on financial markets prevents regressive trading, hence enforcing volatility and subsequent trading. Financial experts indirectly cause more volatility by explaining the market. Expert forecasts, too, provide attributions ("If expert X, too, forecasts that stock Y will rise further, then there might be a really good reason for that"). We find this effect, for instance, in the Investment Dartboard of the *Wall Street Journal*. A study on the years 1988–1990 found evidence for average abnormal returns of 4% and average volume double normal volume for the stocks recommended (Barber & Loeffler, 1993), the positive abnormal return on announcement being partially reversed within 25 trading days. No wonder there seems to be herd behavior even among experts: Analysts' revisions for expected earnings have been found to depend largely on other analysts' revisions (Stickel, 1990, 1991). If we consider variance or volatility, respectively, as risk, shall we then conclude that expert advice increases risk?

In 1995, Robert A. Haugen of the University of California, Irvine, published *The New Finance*. He summarized evidence against the idea of efficient markets that he called a "fantasy" (p. 11). The Capital Asset Pricing Model (CAPM) implies a trade-off between risk and return. According to theory, we can invest in risky (i.e., highly volatile) stocks if there is an acceptable expected return. Yet as Haugen (1995) summarized, as a matter of fact, exactly the opposite is true. In the last 40 to 50 years, the low-volatile stocks have outperformed the high-volatile stocks. This means that low-risk investments produced higher returns than high-risk investments. This is true for a 5-year horizon performance as well as for annualized returns. Haugen commented: "We *want* higher returns on high-volatility stock portfolios, but we don't *get* them because we overreact to the past and ultimately receive low returns in the future" (p. 94).

Moreover, Haugen demonstrated that this countertheoretical phenomenon started at the time portfolio theory was introduced. During the 1930s, 1940s, and through most of the 1950s, the investment strategy of Graham and Dodd's (1934/1951) "Security Analysis" dominated. Graham and Dodd thought expected value to be purely speculative and based stock assessment on its demonstrated performance:

> [T]he analyst's philosophy must still compel him to base his investment valuation on an assumed earning power no larger than the company has already achieved in some year of normal business. This is suggested because, in our

opinion, investment values can be related only to demonstrated performance. (p. 422)

Graham and Dodd's investment strategy was followed by the idea of portfolio selection in terms of risk and expected return. As Fig. 5.4 shows, the correlation between risk and return was valid throughout the 1930s and 1940s and lost its factual basis at the moment when portfolio selection theory was introduced. The reverse of low-risk stocks profitability seems to be an unintended effect of the introduction of the portfolio theory. The theory undermined its own factual basis.

Haugen's conclusions are supported by George Soros, who is both an expert theorizing about financial markets and an expert in this field because he advises one of the world's best performing funds, The Quantum Fund. In his book, *The Alchemy of Finance* (1994), he introduced his concept of *reflexivity* of financial markets. He started with a discussion of the epistemological difference between natural sciences and social sciences. Whereas natural scientists deal with "phenomena that occur independently of what anybody says or thinks about them" (Soros, 1994, p. 32), there is an interaction between the scientists' thinking and the concerned phenomena in social sciences; because the social situation is "contingent" on the partici-

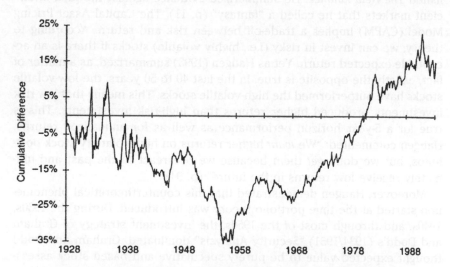

FIG. 5.4. Cumulative difference in return between low-volatile stocks and the standard stock portfolio S&P 500. Until 1958, low-volatile (i.e., low-risk) stocks produced less returns than average stock portfolios. After 1958, according to Haugen, the situation was reversed. *Note.* From *The New Finance* (p. 100) by R. A. Haugen, 1995, Englewood Cliffs, NJ: Prentice-Hall. Copyright 1995 by Prentice-Hall. Reprinted by permission.

pants' decisions (p. 33). Reflexivity, according to Soros, means that two "functions" interact:

- "the participants' efforts to understand the situation" and
- "the impact of their thinking on the real world" (p. 40).

Reflexivity in the stock markets disproves the idea of perfect competition, Soros pointed out. Reflexivity leads to two market distortions. First, markets are "always biased in one direction or another" (p. 49), the reason being self-reinforcing processes in stock market valuations of the kind of the "herd" effect. Second, markets "can influence the events they anticipate" (loc. cit.). Stock market valuations have an impact on the underlying values. There is a direct influence through the transactions triggered by stock prices (e.g., mergers and acquisitions). There are indirect influences on the standing of a company (e.g., credit rating, management credibility).

As we have seen in the Asian crisis of 1997, by increasing the volatility, reflexivity not only leads to instability in financial markets, but also threatens the stability of societies, in particular developing countries, Soros (1998) contended. This is another story.

As to our discussion on experts and forecasting financial markets, we come to the following conclusions:

(a) Financial markets are not efficient in the strong sense that any relevant information will be instantaneously reflected by security prices; thus, investors are still able to earn excess returns by using relevant information (planning information, insider information). Yet financial markets seem to be weak-form efficient, excluding excessive gains from the use of historical information.

(b) We can be skeptical about strong-form experience-based expertise in forecasting financial markets. Forecasting financial markets seems to be limited by a systemic feedback effect: The prediction may have unpredicted counteracting effects on the markets.

(c) Experts may play an important role in transmitting the systemic effects. Providing information and causal attributions (for price changes), experts' advice can increase investor trading behavior.

Having a reasonable hypothesis seems to be the sufficient working base for experts' forecasts in financial markets. In financial markets, information, especially insider information, seems to be more effective than experience-based knowledge. We may call financial experts *blind* in the Kantian sense: They have models and measures, but they lack insight into the real market structures—they lack sufficient perception schooling. The 1980s saw a lot of investment strategies—such as junk-bonds, leverage buy-outs, risk arbitrage—that worked out for a short while and ended in several bankruptcies (e.g., Baily, 1991; Wood, 1988). The systemic pitfalls add to uncertainty about the behavior of financial markets; financial forecasts do not simply try to mirror the unknown future of financial markets, but also influence the markets. In this sense, financial forecasts differ completely from other domains of forecasting: A weather forecast will not influence the weather.

In summary, we can be skeptical about experience-based expertise in forecasting financial markets. Financial markets show systemic effects. There seems to be an intricate recursivity and interdependence of the participants (and their actions). Moreover, expert forecasting is one transmitting factor for volatility: The experts are blind in that they can only try to follow some hypotheses, but lack insight in the complexity that drives the market.

CHAPTER

6

Case Study II: Predicting Climate Change 1988–1997

We leave the field of finance for a while and turn to the problems of environmental protection. As in the case of financial markets, it would be useful to have reliable information on the future development of environmental systems: This is as true for global problems, such as climate change, as it is for defined local problems, such as contaminated sites. In contrast to financial markets, there is also substantial uncertainty about how to understand the present state of environmental systems.

This chapter presents a case study on expertise and experts involved in research on climate changes. It roughly comprises the 10 years between 1988 and 1997, commencing with the foundation of the Intergovernmental Panel on Climate Change (IPCC) in 1988 and ending with the Kyoto Protocol of 1997, through which the industrialized countries bound themselves to specified reductions in greenhouse-gas emissions. The first section focuses on scientific expertise. We see the many ways that uncertainty comes into play and how scientists deal with it. The second section shows how some scientists try to take into account the social and local uncertainties by including citizens and local experts. We find that even when dealing with the global problem of climate change, a kind of lay expert seems required: the local system expert.

6.1 CLIMATE CHANGE AS A MATTER OF SCIENTIFIC CONCERN

The Problem

In 1990 and 1996, the IPCC published two detailed reports on the scientific assessment of climate change. The IPCC was jointly founded by the World Meteorological Organisation (WMO) and the United Nations Environment

Programme (UNEP) in 1988. The main issue of the reports was information on the causes of climate change, especially global warming caused by the so-called *greenhouse gases* such as carbon dioxide (CO_2).

> There is concern that human activities may be inadvertently changing the climate of the globe through the enhanced greenhouse effect, by past and continuing emissions of carbon dioxide and other gases which will cause the temperature of the Earth's surface to increase—popularly termed the "global warming." If this occurs, consequent changes may have a significant impact on society. (IPCC, 1990, p. xiii)

On the basis of a scientific assessment, the reports were to present "current knowledge regarding predictions of climate change (including sea level rise and the effects on ecosystems) over the next century, the timing of changes together with an assessment of the uncertainties associated with these predictions." *Climate* is usually defined to be average weather, described in terms of the mean and other statistical quantities that measure variability over a period of time and over a certain geographical region (Trenberth, Houghton, & Meira Filho, 1996, p. 55).

Climate change has to be regarded as changes to the climate system. Defined by the UN Framework Convention on Climate Change, the climate system is the "totality of the atmosphere, hydrosphere, biosphere and geosphere and their interactions" (UNEP/WMO, n.d., p. 5). Thus, the climate system includes not only the atmosphere, but also the oceans, ice, the land, and, last but not least, the earth's biomass such as forests (see Fig. 6.1).

The most important component influencing the climate system from outside is the sun. The earth intercepts solar radiation; about a third of it is reflected, and the rest is absorbed by various components of the climate system. The total energy absorbed from radiation is balanced by outgoing radiation from the earth and the atmosphere. This terrestrial radiation takes the form of invisible long-wave infrared radiation. There is a phenomenon that prevents a great deal of radiation from being reemitted by the earth—the so-called *greenhouse effect*. Short-wave radiation from the sun passes through the clear atmosphere relatively unimpeded, but long-wave terrestrial radiation emitted by the warm surface of the earth is partially absorbed and then reemitted by a number of trace gases in the cooler atmosphere above. Because the energy absorbed is also reemitted downward, the earth's surface and the lower atmosphere will be warmer than they would be without the greenhouse trace gases. The main natural greenhouse gases are water vapor, carbon dioxide (CO_2), and methane (CH_4). Also small particles in the atmosphere, so-called *aerosols*, can reflect and absorb radiation and thus affect the climate.

The concentration of carbon dioxide in the atmosphere and the temperature change have strongly correlated over the last 160,000 years (studied by

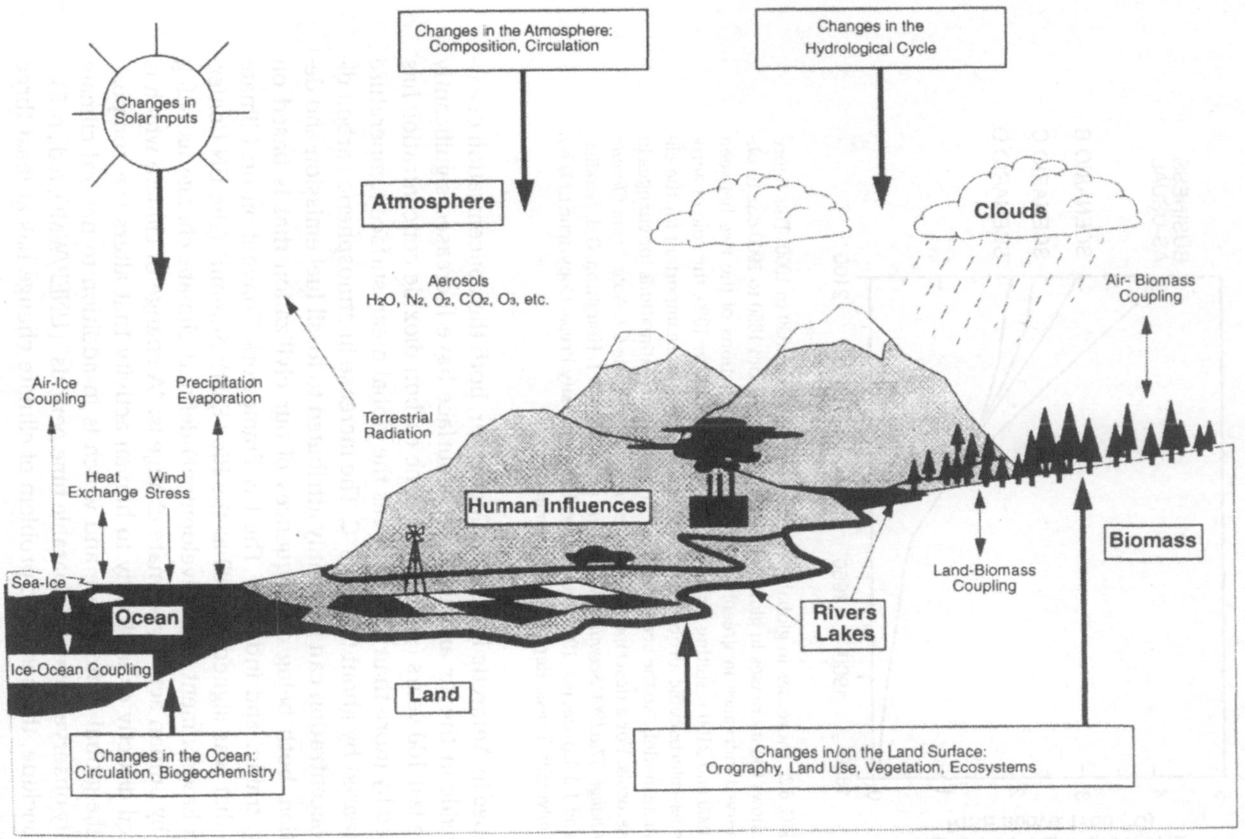

FIG. 6.1. Schematic view of the global climate systems. This figure shows the climate systems (bold), their processes and interactions (thin arrows), and some aspects that may change (bold arrows). *Note.* From *Climate Change 1995: The Science of Climate Change* (p. 55) by T. Houghton et al., 1996, Cambridge, UK: University Press. Copyright 1996 by University Press. Reprinted by permission.

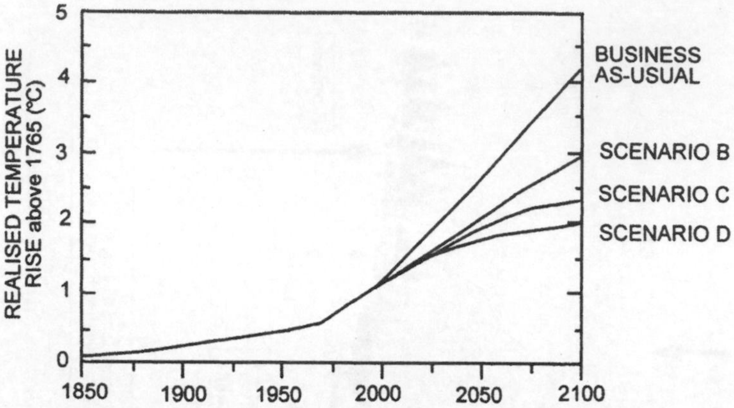

FIG. 6.2. Increases in global mean temperature from 1850 to 1900. This figure shows the increases in global mean temperature from 1850 to 1900 due to observed increases in greenhouse gases, and predictions of the rise between 1900 and 2100 resulting from different scenarios. After 1990, the role of aerosols—microscopic airborne particles—has become more important for the climate models, so the scenarios were amended with estimations for changes in aerosols. For a description of the scenarios, see Table 6.1. *Note.* From *Climate Change: The IPCC Scientific Assessment* (p. xxiii) by J. T. Houghton, G. J. Jenkins, and J. J. Ephraums, 1990, Cambridge, UK: University Press. Copyright 1990 by University Press. Reprinted by permission.

air trapped in Antarctic ice cores). Moreover, both the concentration of carbon dioxide in the air and the earth's surface have increased significantly over the last 100 years (see Fig. 6.2). The carbon dioxide concentration has increased by more than 25%, whereas the global mean surface temperature has increased by about 0.3 to 0.6°C. The increase in atmospheric carbon dioxide concentration can be mainly attributed to fossil fuel emission and deforestation, both being consequences of our civilization that is based on housing, traffic, and industry. The UN Framework Convention on Climate Change that was signed in 1992 at the Rio "Earth Summit" (the UN Conference on Environment and Development) defined climate change as being caused by human activity; climate change is: "A change of climate which is attributed directly or indirectly to human activity that alters the composition of the global atmosphere and which is in addition to natural climate variability observed over comparable time periods" (UNEP/WMO, n.d., p. 5).

To conclude, the scientific problem of climate change has at least three aspects:

- understanding the structure and dynamics of the climate system in general,
- assessing the contribution of human activities, and

- predicting the future development of the climate.

The most sophisticated models used to simulate climate change are so-called *general circulation models* (GCMs). GCMs are physically based models that describe "the atmospheric and oceanic dynamics and physics" and are also based "upon empirical relationships, and their depiction as mathematical equations" (IPCC, 1996, p. 31). The resolution of a GCM is not very deep: Atmospheric GCMs normally depict the climate system in parts that are 250 km in the horizontal and 1 km in the vertical. Thus, the accuracy of modeling with GCMs still is limited.

Scenario Construction

The IPCC report of 1990 communicated the results in four scenarios of climate policies for the time until the year 2100. Scenario A depicts business as usual:

> [T]he energy supply is coal intensive and on the demand side only modest efficiency increases are achieved. Carbon monoxide controls are modest, deforestation continues until the tropical forests are depleted and agricultural emissions of methane and nitrous oxide are uncontrolled [. . .]. (IPCC, 1990, p. xxxiv)

The other three scenarios depict increases in efficiency in the use of carbon fuels, carbon monoxide controls, halted deforestation, and a shift toward renewables and nuclear energy (see Table 6.1). Under the scenario business-as-usual, the IPCC predicted a rate of increase in global mean temperature during the next century "of about 0.3°C per decade" with an "uncertainty range of 0.2°C to 0.5°C per decade" (p. xi). The Climate Change 1995 report (IPCC, 1996) had to correct this figure downward. Now the best estimate was about 0.2°C per decade. "This is due primarily to lower emission scenarios [. . .], the inclusion of the cooling effect of sulfate aerosols, and improvements in the treatment of the carbon cycle" (1996, p. 6)—the attentive reader of the reports may also notice a change in terminology: Whereas the 1990 report speaks of "predicting" climate change, the 1995 report speaks of "projecting" it (see IPCC, 1996, p. 31; 1990, p. xxv).

We take a short look at scenarios and scenario construction from three points of view: methodology, cognitive psychology, and professional work. We start with the method. The scenario method was introduced by Herman Kahn in the late 1960s as a tool for strategic planning. In his book (with Wiener), *The Year 2000*, he explained what scenarios are for:

> Scenarios are hypothetical sequences of events constructed for the purpose of focusing attention on causal processes and decision-points. They answer

TABLE 6.1
IPCC Climate Change Scenarios as Communicated in the IPCC 1990 Report

	CO_2				Methane & Nitrous Oxide	CFC	
	Energy Mix	*Supply: Efficiency*	*Carbon Monoxide-Controls*	*Deforestation*	*Agriculture**	*Montreal Protocol***	*Timing*
A "Business as usual"	Coal	–	Modest	+	+	Partially implemented	
B "Low emissions"	Natural gas	+	+	–	?	+	
C "Control policies"	Renewables & nuclear power	+	+	–	–	CFC phase out	> 2050
D "Accelerated policies"	Renewables & nuclear power	+	+	–	–	CFC phase out	< 2050

Note. Source: IPCC (1990, p. xxxiv). Only some of the parameters used for scenario construction are communicated (for details, cf. den Elzen, 1994; IPCC, 1992). The construction was undertaken by a U.S.–Dutch expert group (see den Elzen, 1994). The scenarios are constructed by varying the state of the different parameters for greenhouse-gas emissions. To some extent, the scenarios build on one another: Scenario B is "better" than A, as C is "better" than B, and so on; Scenario C contains the same preferable regulations as Scenario B and some more (such as the use of renewables). Scenario D has an additional time limit. The table shows only those specifications that were explicitly made; for instance, we can assume that Scenarios C and D have carbon monoxide (CO) controls. The scenarios were revised in 1992 (IPCC, 1992) and evaluated in 1994 (IPCC, 1995).

*There are methane (CH_4) emissions form rice paddies and from enteric fermentation in cows. There are emissions of nitrous oxide (mainly N_2O) from fertilizer use in agriculture.

**Regarding chlorofluorocarbons (CFC), there had already been an international agreement on regulation formulated in the Montreal Protocol of 1987; this was in the context of regimes against stratospheric ozone depletion (see e.g., UNEP, 1996).

> two kinds of questions: (1) Precisely how might some hypothetical situation come about, step by step? and (2) What alternatives exist, for each actor, at each step, for preventing, diverting, or facilitating the process. (Kahn & Wiener, 1967, p. 6)

The scenario method is used in corporate planning (e.g., with Shell Corporation; see Ringland, 1998), as well as in political planning, for example, as to energy consumption (e.g., Prognos AG, 1996). There is also use of scenarios in team development (e.g., Fahey & Randall, 1998) and team teaching (e.g., Scholz & Tietje, in press).

From the point of view of cognitive psychology, the construction of scenarios is an instance of mental simulation. This is true for someone who constructs the scenarios as well as particularly for someone who has to understand–to reconstruct–them. We can divide the process of scenario construction into four steps (see Fig. 6.3). The first step consists in the activation of knowledge about the problem; the second step leads from knowledge of the problem to the construction of a mental model; the third step consists of inferences from the model; in the last step, we have to select relevant inferences to arrive at useful scenarios. We discuss the four steps in the light of two cognitive tasks: problem representation and delimiting relevance.

1. The cognitive representation of the relevant domain properties and the problem aspects are mainly based on the constitution of a mental model (Johnson-Laird, 1983). Mental models contain causal knowledge of this type: "If parameter X changes, then parameter Y will change, too." A mental model for assessing climate change would include a model of the climate system (as in Fig. 6.1), physical laws, and knowledge about relevant parameters (e.g., the current amount of CO_2 emissions). The mental model is the central mechanism for the cognitive simulation (e.g., the simulation of future problem development).

2. The selection of inferences delimits the relevant from the irrelevant properties. The IPCC (1990) scenarios are based on the variation of several key parameters that supposedly influence emissions of greenhouse gases, including: population growth, economic growth, deforestation, end-use efficiency of energy uses (see den Elzen, 1994, chap 6; IPCC, 1992). The scenarios, however, are communicated as policy scenarios that focus on politically relevant joints–decision points–such as carbon monoxide controls and the implementation of the Montreal Protocol that regulates CFC emissions (as in Table 6.1). These joints are dramatic events "that are low in redundancy but high in causal significance" (Kahneman & Tversky, 1982, p. 207). The use of joints results in a "tendency to underestimate the likelihood of events that are produced by slow and incremental change" (loc. cit.). The problem the IPCC scenarios are facing is complementary: They have to plausibly explain

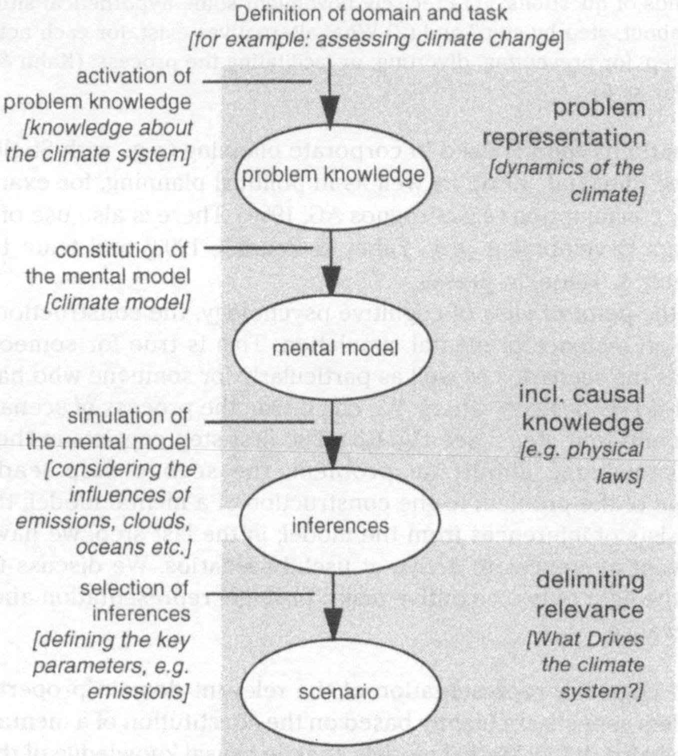

FIG. 6.3. Mental scenario construction from a psychological point of view. The examples and the shaded spaces are added. *Note.* From "The Labyrinth of Experts' Minds: Some Reasoning Strategies and Their Pitfalls" by H. Jungermann and M. Thüring, 1988, *Annals of Operations Research, 16*, Fig. 1. Copyright 1988 by *Annals of Operations Research*. Adapted by permission.

the incremental change of climate by linking it to dramatic joints, especially in the field of national and international politics.

Let us take a closer look at scenario construction as part of professional work or, more precisely, as professional inference. Abbott mentioned the aspects of construction and exclusion. Construction and exclusion are relevant to the IPCC scenarios. The aspect of *construction* is obvious; each scenario results from some kind of construction: Three scenarios (A to C) "were designed in such a way that they envisaged a doubling of the CO_2-equivalent concentrations in the years 2030, 2060, and subsequently in 2090" (den Elzen, 1994, p. 48). In addition, "the expert group was asked to develop an additional fourth scenario [D] that would lead to stabilization of the CO_2-equivalent concentration at a level below a doubling of pre-industrial atmospheric CO_2" (loc. cit.).

The aspect of inference by exclusion is less evident. If we consider the number of parameters that have to be varied, 16 in the description of the scenarios (den Elzen, 1994), we find a minimum of 65,536 possible scenarios ($=2^{16}$), assuming that every parameter can change in one way. Therefore, the interpretative *Leistung* of the IPCC scenario experts not only consists in describing what might happen, but also in selecting relevant scenarios. For example, the scenario description in Table 6.1 takes into account the different global warming potentials (GWP) of greenhouse gases: For CO_2 that is high in GWP, there are more parameters than for methane or CFC that are low in GWP.

The global warming potential (GWP) is a policy-adaptable coined measure that emphasizes the role of CO_2: "the warming effect of an emission of 1 kg of each gas relative to that of CO_2" (IPCC, 1990, p. xxi). On a 100 year horizon, CO_2 has a GWP of 1, CH_4 of 21 (changed to 11 according to IPCC, 1992, p. 15), Nitrous oxide of 290 (270 in 1992).

To conclude, we can state that scenarios combine prediction and planning. Planning presupposes to some extent freedom and causality. If we had "total causation" (Putnam, 1983), nothing would be left for planning and regulation. Similarly, total uncertainty would render both prediction and planning impossible. Scenarios provide hypothetical knowledge. They predict the future under some hypothetical conditions: If no carbon dioxide controls are implemented before year X, then the effect on climate change will be Y. The IPCC scenarios try to link models of the climate system to the "many social and economic uncertainties" (den Elzen, 1994, p. 57). The causal inferences the prediction is based on are derived from the current scientific climate model as provided by the general circulation models (GCMs). The planning information is left open or, respectively, left to politics.

Scientists and Uncertainties

A threat to IPCC scenarios are the many uncertainties that remain. From the point of view of IPCC, there are two sources of uncertainties in the detection of climate change and the attribution of causes:

- uncertainties in model projections of anthropogenic change, such as "inadequate representation of feedbacks" or "signal estimation problems" (IPCC, 1996, pp. 416–417);
- uncertainties in estimating natural variability (p. 418).

Under the heading "Narrowing the Uncertainties," the 1990 IPCC report also presented a detailed research timetable for overcoming the uncertainties

(McBean & McCarthy, 1990). The heading was changed into "Advancing our Understanding" in the 1995 report (McBean, Liss, & Schneider, 1996).

However, as Pahl-Wostl (1995) argued, the problem in predicting climate change is more fundamental and cannot be solved by improving the model projections and parameter estimations. She maintained that the uncertainties involved have to be attributed to

- insufficient knowledge, as well as
- characteristics "inherent in the dynamics of nonlinear systems" (p. 197).

According to Pahl-Wostl, the logic of the research conducted by IPCC is driven by linear thinking that is misleading when it comes to understanding the nonlinear climate system. Pahl-Wostl wrote:

> Linear thinking is implicit in the belief that the anthropogenically imposed effect of greenhouse gases can be separated from the natural endogenous dynamics of climate. This assumption is a necessity for the detection and the prediction of any greenhouse effect. The increasing knowledge about the behavior of nonlinear systems does not lend much support to such an assumption.

The critical point is the possibility of sudden jumps in system behavior. Pahl-Wostl showed how these jumps can be simulated using the so-called *Lorenz model.* She argued: The sudden jumps in model behavior derive from the presence of a "critical parameter that determines the threshold to chaos" (p. 200). The example given by Pahl-Wostl (1995) is the sudden shift in the North Pacific climate that suddenly occurred in the mid-1970s (see Fig. 6.4).

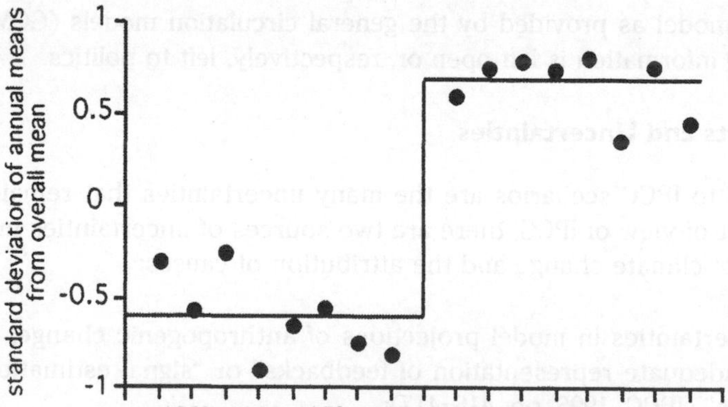

FIG. 6.4. Sudden climate change in the North Pacific region in 1976 shown as an index of 40 environmental variables. *Note.* From *The Dynamic Nature of Ecosystems* (p. 202) by C. Pahl-Wostl, 1995, Chichester: Wiley. Copyright 1995 by Wiley. Reprinted by permission. See also Kerr (1993).

> This shift is related to the El Niño phenomenon: "The climate shift that gripped a good part of the Northern Hemisphere turns out to have been a new side to a familiar problem–El Niño. In essence the on-off seesaw of El Niño, which ordinarily floods the eastern Pacific with warm water every 3 or 4 years, got stuck part way 'on' for 10 years straight, prolonging the wild weather that El Niño often brings." (Kerr, 1992, p. 1508)

Pahl-Wostl (1995) concluded that in the climate system as well as in nonlinear systems in general:

- Linear extrapolations are not feasible.
- Predictions of threshold effects are exceedingly difficult.
- Relevant factors may not be recognized as such due to their insignificance in an undisturbed situation.
- Clear cause-effect relationships are virtually nonexistent. Effects are contingent on state and context. (p. 206)

Therefore, Pahl-Wostl argued the necessity of developing a "new type of knowledge" to deal with "uncertainties that obviously cannot be avoided but only neglected" (p. 208).

To pay tribute to the scientific discussion, we should be aware that Pahl-Wostl's argumentation is not unanimously taken for granted. For instance, in a personal comment in 1998, Atsumu Ohmura, a reviewer of the 1990 IPCC report, expressed strong reservation about Pahl-Wostl's interpretations. In particular, the sudden shift in the North Pacific climate "has nothing to do with linearity and nonlinearity." If Pahl-Wostl is right, there would be no use for linear extrapolations. This would also affect important climate change parameters such as climate sensitivity that IPCC defined as the "long term change in surface air temperature following a doubling of carbon dioxide" (IPCC, 1990, p. xxv). The best estimate is that the global mean surface temperature will–linearly–increase by 2.5°C if the atmospheric CO_2 concentration doubles.

In a thorough study in 1995, Morgan and Keith asked 16 high-ranking U.S. climate scientists about climate sensitivity and other climate policy-relevant factors. They found both an astonishing homogeneity in predicted climate sensitivity as well as a generally "greater disagreement than we believe is usually conveyed in scientific consensus documents. [. . .] We find there is almost no agreement about the effect of climate change on policy-relevant factors such as changes in precipitation over land and various forms of interannual variability" (1995, p. 468). The scientific disagreement cannot really surprise and is not exclusive to climate change problems. More interesting is the relative agreement on climate sensitivity. Climate sensitivity is used as a benchmark when comparing climate models. Thus, climate sensitivity is a central measure of modeling climate change. The overall mean of the experts' estimation for climate sensitivity is 2.8°C, with a range of 0.3 to 4.7°C. The mean is not far from the best 1990 IPCC estimate of 2.5°C, but the range is outside the 1990 accepted range of 1.5 to 4.5°C.

In the 1995 Morgan and Keith study, the climate sensitivity estimates are of central importance. If we look at the data closely, we find a narrow correlation between the climate sensitivity estimates and other parameters.

> In particular, there is a cooling effect of aerosols—microscopic airborne particles that have been one reason, among others, for reducing the global mean surface temperature predictions (IPCC, 1996, p. 6). The climate sensitivity estimates in the 1995 Morgan and Keith study significantly correlate with the estimates for aerosol influence—namely, the estimates for the "contribution today of anthropogenic aerosols to radiative forcing" (radiative forcing means the impact on climate change). The higher an expert estimates climate sensitivity, the higher he also estimates the reduction of radiative forcing through aerosols. Thus, effects of aerosols today serve as regulators for high climate sensitivity estimates. We can understand the parameter *climate sensitivity* as a kind of consensus parameter within the scientific community of climate change research.

Despite relative agreement on the figures, almost all of the experts in the Morgan–Keith study consider the uncertainty in the climate sensitivity as critical: "Most judged uncertainty in the large-scale sensitivity as the dominant factor in the total uncertainty, 9 of 10 ≥ 70%. In addition to three 'don't knows,' qualitative responses included words like 'substantial' and 'could be dominant' " (Morgan & Keith, 1995, p. 472). Thus, it would be interesting to see whether climate sensitivity proves to be a "matter of reality"—and not only as point of consensus within the thought collective IPCC.

To conclude, we turn back to the link of climate change research to politics. IPCC is a successful example of bridging science and politics on an international level. First, IPCC creates some scientific consensus that can be communicated as the scientific knowledge base. Second, the IPCC scenarios directly connect scientific modeling to political decision making. Thus, the IPCC scenarios are a useful means to promote the interests of the scientific community, in particular as to funding new research. However, there are also doubts that more research results in better politics. For example, Sonja Boehmer-Christiansen (1994a), who investigated the history of the climate change protection policy, formulated three fundamental questions that concern the use of general circulation models (GCMs), but also relate to the IPCC climate change research in general:

> First, should climate be analyzed within the paradigm of change and stability or would another paradigm be more appropriate? What constitutes climate "change?" [. . .]

TABLE 6.2
Climate Sensitivity

	Climate Research Board (1979)	*Dickinson (1986)*	*1990, IPCC*	*Morgan & Keith (1995)*	*1995 IPCC*
Climate sensitivity (the long-term change in surface air temperature following a doubling of carbon dioxide)	≈3.0°C (range 1.5–4.5°C)	3.5°C (accepted range 1.5–5.5°C)	2.5°C (accepted range 1.5–4.5°C)	Average 2.8°C range 0.3–4.7°C	Cited studies (293) Ø 4.3°C range 2.8–5.2°C (accepted range 1.5–4.5°C)
Change in global mean temperature by 2100		3.5°C (range 1.5–5.5°C)	3°C range 0.9°C to ≈3.5°C	above 0.5°C below 3.6°C*	2°C range 1°C to 3.5°C

Note. From *Global Environmental Change* (p. 44) by M. den Elzen, 1994, Utrecht: International Books. Copyright 1994 by International Books. Adapted by permission.

*Figures are estimated from Morgan and Keith (1995, Fig. 5, p. 473) and refer to the climate system's response to a linear concentration ramp that leads to a doubling of CO_2 over the next 50 years. Two experts saw a possibility of falling temperatures (by 2100 and 2150, respectively) after initial warming (see Morgan & Keith, 1995, p. 472).

> Second, can anthropogenic impacts in practice be isolated from natural ones and predicted with confidence for hundreds of years into the future?
>
> Third, will more scientific knowledge as such (that is the more precise diagnosis of the problem) in fact and virtually automatically generate better policy? (p. 141).

Thus, we not only have to discuss the validity of predictions in environmental issues, but also their role in connection to politics.

6.2 CLIMATE CHANGE AS A MATTER FOR PUBLIC-SCIENCE DIALOGUE

Integrated Assessment

We have seen that IPCC tries to overcome social and economic uncertainties by constructing scenarios. The Dutch National Institute of Public Health and the Environment (RIVM), in coordination with IPCC, developed a more encompassing approach to handle uncertainties—called *integrated assessment.* Integrated assessments are frameworks that combine knowledge from a wide variety of disciplines. These efforts address three goals:

(a) coordinated exploration of possible future trajectories of human and environmental systems,
(b) development of insights into key questions of policy formation, and
(c) prioritization of research needs (Kasemir, Van Asselt, Dürrenberger, & Jaeger, 1999).

Rotmans and Van Asselt (1996) who, among others, developed and promoted the integrated assessment approach added: "(1) it [integrated assessment] should have added value compared to assessment based on a single disciplinary [approach]; and (2) it should provide useful information to decision makers" (p. 121).

Van Asselt and Rotmans argued that the IPCC assessment is characterized by a hierarchical perspective. According to them, perspectives can be characterized by two dimensions: "(1) world view, i.e. a coherent conception of how the world functions, and (2) management style, i.e. signifying policy preferences and strategies" (loc. cit., p. 125). The hierarchical perspective involves a specific myth of nature: According to that perspective, nature is tolerant within certain limits, allowing for some "perversion." The perspective is an ideal typus in the tradition of cultural theory (Thompson, Ellis, & Wildavsky, 1990). Figure 6.5 depicts two further perspectives: an egalitarian one and an individualist perspective. The egalitarian perspective is the most pessimistic one. It supposes that minor changes in the cli-

	hierarchist	**egalitarian**	**individualist**
myth of nature*	perverse/tolerant: nature is tolerant within certain limits	ephemeral: minor changes disproportion-ately influence the system behavior	benign: ecosystems being resilient, nature will find (new) balances
	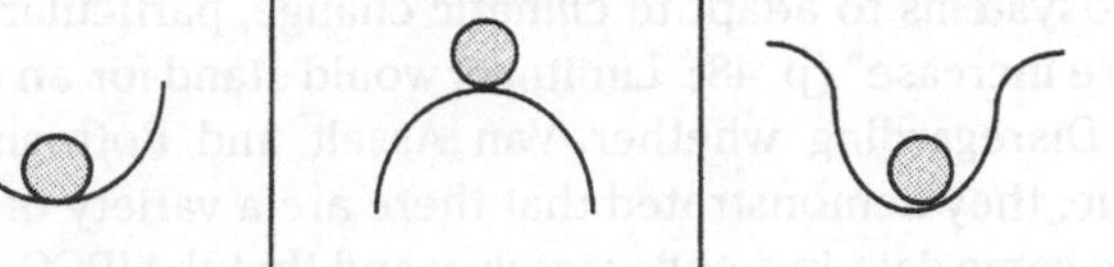		
migration of ecosystems**	neglected	amplifying effect	dampening effect
climate sensitivity**	2.5 ° C	5.5 ° C	1.5° C

FIG. 6.5. Myths of nature according to the integrated assessment approach (see Van Asselt & Rotmans, 1995, 1996). The definition of the perspectives is based on Thompson et al. (1990). The drawings illustrate the myths of nature. They show a ball on a bent surface, symbolizing the different views of the changeable state of nature.

*From *Uncertainty in Integrated Assessment Modelling* (pp. 16, 60) by M. Van Asselt and J. Rotmans, 1995, Bilthoven: National Institute of Public Health and the Environment. Copyright 1995 by NIPHE. Reprinted by permission.

**From "Uncertainty in Perspective" by M. Van Asselt and R. Rotmans, 1996, *Global Environmental Change*, *6*, p. 128. Copyright 1996 by Elsevier. Reprinted by permission.

mate system will have a disproportional impact on the system's behavior. The best estimate for climate sensitivity, according to this perspective, is about 5.5°C and exceeds the estimations that result from other perspectives. The third perspective is the most optimistic one. It assumes that ecosystems will adapt. In particular, the possible migration of ecosystems will have a dampening effect—an interpretation that is totally different than the one from the egalitarian perspective. The figures in Fig. 6.5 symbolize the different myths of nature: an unstable and ephemeral system from the egalitarian perspective, and a stable and benign system from the individualist perspective. The naming of the perspectives results from differences in policy style: The hierarchical perspective stresses the need for a sound scientific base of policymaking; the egalitarian perspective stresses the need for joint public efforts, including nongovernmental organizations; and, according to the individualist perspective, no public climate policy is necessary at all.

An example for differences in scientific views on climate change is provided by Nordhaus (1994), who reported on interviews with social and natural scientists. He found what he called a *frontier view* and the *limitniks*. The frontier view expressed the belief that ecosystems would adapt (individualist perspective) and was mainly supported by economists. For example, an economist explained that, "in his view energy and brain power are the only limits to growth in the long run" (p. 49). The limitniks mainly consisted of natural scientists who "voiced deep concern about the ability of natural ecosystems to adapt to climatic change, particularly for the large temperature increase" (p. 48). Limitniks would stand for an egalitarian perspective.

Disregarding whether Van Asselt and Rotman's (1995) description is true, they demonstrated that there are a variety of possibilities to interpret the same data in a coherent way and that the IPCC view is not exclusive. For instance, there are several good reasons for different best estimates for climate sensitivity depending on the underlying perspective that gives way to different model routes (i.e., different ways of setting the model parameters; Van Asselt & Rotmans, 1995, p. 18).

The integrated assessment approach developed a methodology to analyze uncertainties that is based on an analysis of perspectives. The procedure is similar to the IPCC procedure used to define scenarios. The IPCC defines policy scenarios in terms of degrees of reducing greenhouse gases, whereas integrated assessment takes the cultural perspectives of different climate policies into account. Therefore, from the point of view of integrated assessment, the distinction between scientific uncertainties on the one hand and social and economic uncertainties on the other hand is indeed crucial to understanding and assessing uncertainties:

1. Scientific uncertainties: Those occurring in the environmental subsystem which arise from the degree of unpredictability of global environmental change processes.
2. Social and economic uncertainties: Those occurring in the human subsystem which arise from the degree of unpredictability of future geopolitical, socioeconomical and demographic evolution (loc. cit., p. 11).

Aiming to overcome these uncertainties and providing useful information to decision makers, integrated assessment takes into account the perspective of the public. From the point of view of integrated assessment, this means changing the interaction among science, public, and decision makers toward a dialogue among the groups (see Fig. 6.6):

> The traditional view on [of] the role of science can be described as: Science finds clear-cut descriptions of, and answers to, problems, and informs decision makers in public policy and business about these results. The decision makers then take the necessary measures and feed them into the policy process. This

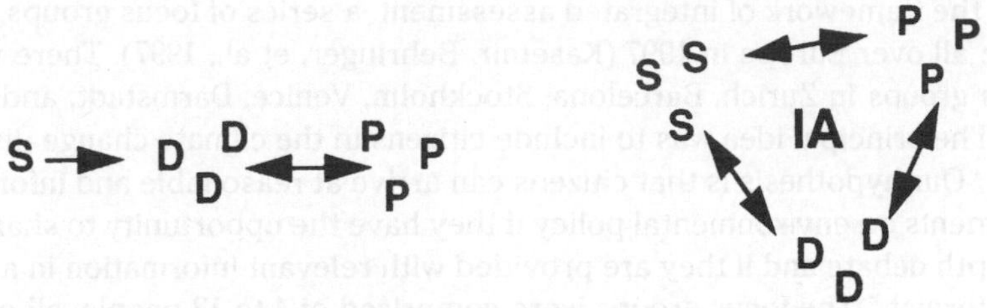

FIG. 6.6. Views of science–public interaction according to the integrated assessment approach. On the left side is a predominant view: Science (S) provides data for decision makers (D) who interact with the public (P). The integrated assessment view of science–public interaction (right side): Science, decision makers, and the public interact in building an integrated assessment. *Note.* From "Integrated Assessment of Sustainable Development: Multiple Perspectives in Interaction" by B. Kasemir, M. Van Asselt, G. Dürrenberger, and C. C. Jaeger, 1999, *International Journal of Environment and Pollution, 11*, Figs. 1 and 2. Copyright 1999 by *IJEP*. Reprinted by permission.

role is most memorably captured in the phrase "speaking truth to power" (Kasemir et al., 1999, p. 409; see also Price, 1965).

In contrast to traditional views of knowledge transfer from science, integrated assessment is based on *informed judgments*: "Procedures to arrive at an informed judgment on different courses of action with regard to complex environmental problems. The information required refers to natural and social scientific knowledge, and also concerns the relevant decision making process."

Focus Groups, Lay Experts, and System Experts

One procedure for the integrated assessment of complex environmental problems—also in climate change—are focus groups. A focus group is a "guided group discussion that is focused on a specific topic" (Dürrenberger, 1997, p. 6). Focus groups are common in market research (e.g., Krueger, 1988). They mostly consist of a homogenous segment of consumers (housewives, the new rich, etc.) or a balanced selection of these segments, respectively. Discussion in the groups is supported by a skilled moderator. Group discussions that combine public, science, and decision makers are not a new procedure and not exclusive to environmental matters. They are also commonly found in mediation processes, where the interest of different social groups have to be negotiated to prevent overt conflicts (e.g., Wates, 1996). As to environmental issues, we find round table discussions on energy and urban problems as well as scientifically focused groups (e.g., the area development negotiations; Scholz & Tietje, in press).

In the framework of integrated assessment, a series of focus groups took place all over Europe in 1997 (Kasemir, Behringer, et al., 1997). There were focus groups in Zurich, Barcelona, Stockholm, Venice, Darmstadt, and Athens. The principal idea was to include citizens in the climate change discussion: "Our hypothesis is that citizens can arrive at reasonable and informed judgments on environmental policy if they have the opportunity to share an in-depth debate and if they are provided with relevant information in a suitable format." The focus groups were comprised of 4 to 12 people, all of the invited participants being nonexperts regarding climatology. Each group met several times and discussed the regional and personal importance of climate change, sustainability, and urban lifestyles. The groups were guided by one or more moderators. The integrated assessment was brought about through:

- fact sheets or working papers that informed about global warming and basics of climatology;
- computer models for integrated assessment such as *targets* (Rotmans & de Vries, 1997) that the groups tested and applied to their specific discussions; and
- consensus reports or citizen reports that tried to identify goals and policies for regional measures to prevent climate change.

There are two interpretations for the necessity of integrating the perspective of the public—namely, a political interpretation and an ecological one. The political interpretation stresses the importance of democratic processes. This is sometimes the point of view of integrated assessment models (e.g., Kasemir & Jaeger, 1997). According to the integrated assessment approach, modeling and data collection seem to be the task of science. The methods of handling uncertainty are scientific ones. The public enters the discussion when it comes to finding an interpretation and possible problem solutions.

In an ecological interpretation, ecosystems must be seen from a relational perspective: "Organisms have no identity in isolation but gain their identity only in the context of their environment" (Pahl-Wostl, 1995, p. 191). Moreover, ecosystems have different degrees of autonomy. As to dynamic systems, a system is called *autonomous* if: "The defining equations do not include any external forcing terms. The dynamics of autonomous systems are determined exclusively by the state variables and the internal processes connecting them" (loc. cit., p. 194). In this strict sense, no ecosystem is autonomous because they are open systems that exchange energy and matters. As Pahl-Wostl added, "there may be varying degrees of being autonomous beyond the simple distinction between yes or no" (loc. cit., p. 194).

Human beings can be part of certain ecosystems, too. Thus, to understand the functioning of the system, we have to understand the views and behavior of the persons involved. This is especially true in urban planning (e.g., Buckingham-Hatfield & Percy, 1999; Park, Burgess, & McKenzie, 1967; The Prince of Wales's Urban Task Force, 1997). Moreover, if the system is somewhat autonomous, such as some suburbs of big cities, the persons living there are genuine sources of information, which is also necessary to discover and evaluate a scientific view of the system—an approach sometimes referred to as *transdisciplinarity* (Thompson, Klein, et al., 2000) or Mode 2 research (Gibbons et al., 1994).

Whatever interpretation of the necessary integration of the perspective of the public we may prefer, we are concerned with a special form of experts: lay experts or system experts, respectively, who provide local knowledge. As to environmental matters, scientific experts have no exclusive power in defining the problem space and providing analyses. Because environmental matters are intrinsically connected to human action, it may be wise for policy-oriented research to integrate it from the beginning. The kind of knowledge on which the specific *Leistung* of lay or system experts is based is:

(a) *local planning knowledge*: What can be done in this particular town? Who has to be involved? How do you raise funds and support for local environmental policies?

(b) *local environmental system knowledge*: What is the history of the local environment? What are its peculiarities? These questions not only pertain to issues of contaminated soils or area development, but also to climate change policy: CO_2 emissions from housing and agricultural methane emissions differ according to the local housing situation and agriculture.

The term *system expert* has been created in the context of the Swiss ETH-UNS case studies; these are projects for environmental planning (e.g., concerning the reintegration of urban industrial areas; see Mieg, 1996; Scholz, Bösch, et al., 1997). In the context of these projects, the system experts report on the case and represent the local network of professionals concerned with the particular local environment (e.g., as architects or geologists; see Mieg, 2000). In the remainder of this book, we use the term *system experts* rather than *lay experts* or *locality experts*. System experts are experts on a local system—maybe a local environment, maybe a community. System experts are experts through interaction of the "The expert"-form; they are experts relative to a problem and other persons who are concerned with the problem. Their knowledge is concrete and need not be based on an academic education or scientific background. Rather, it is based on local experience. System experts interpret the local system for other experts or decision makers.

We can also interpret the appearance of environmental system experts from the expertise perspective—apart from the political dimension. Environmental expertise is often based on scientific knowledge about the problem, but not on practical experience in dealing with the problem, in particular in case of the climate change issue. Although meteorologists, according to Shanteau, fall within the group of experts who perform well (see Table 2.3), climate change research could hardly be classified because it has not, so far, been possible to assess its performance. In general, scientific knowledge about environmental problems does not necessarily constitute a strong-form expertise in tackling these problems. Moreover, there is no fully fledged environmental profession yet, at least not in Europe (see Mieg, 1998). Thus, a purely scientific approach to concrete environmental problems—as well as a traditional engineering approach—leaves room for the use of experts with experience in a particular environmental system.

To conclude, integrated assessment has not become the view supported by the Kyoto Protocol. This protocol was adopted by the third session of the conference of the parties to the UN Framework Convention on Climate Change (UNFCCC) that took place in Kyoto, Japan, in December 1997 (see Grubb, 1999). The Kyoto Protocol specifies national targets for the reduction of the emission of greenhouse gases, in particular carbon dioxide (although adopted, so far the Kyoto Protocol has not come into force). Like the UNFCCC of 1992 (see UNEP/WMO, n.d.), the Kyoto Protocol requires national inventories of anthropogenic emissions. Unlike the UNFCCC, the Kyoto Protocol relies on expert review teams that should provide a "thorough and comprehensive technical assessment of all aspects of the implementation" (Art. 8; see Grubb, 1999, p. 288). This does not mean that the Kyoto Protocol solely relies on scientific expertise because experts are selected from several parties, including the contracting states and intergovernmental organizations. In the next chapter, we take a closer look at the various types of services and roles connected to experts.

In summary, predicting change in environmental systems has to deal with many uncertainties. The International Panel on Climate Change (IPCC) assessment of climate change is communicated in the form of scenarios that predict global change relative to environmental policies. The integrated assessment approach claims that cultural policy perspectives also have to be taken into account, resulting in the idea of focus groups combining science, decision makers, and the public. As a matter of fact, persons outside of science act as system experts when it comes to analyzing local environments in urban as well as rural systems.

CHAPTER

7

Conclusions for the Conceptualization of Expertise in Context: Types of Experts, Uncertainty, and Insecurity

Having discussed experts in financial markets and climate change, we now turn back to "The expert"-interaction and formulate some conclusions. We find that different uses result for different types of experts such as scientists and professionals. The starting point is the fact that not every kind of uncertainty creates a demand for cognitive expertise. We have to separate the epistemic question (what do we know?) from the pragmatic question (what shall we do?). This chapter introduces a typology of experts depending on the knowledge they administer and the use we can make of them. The typology of experts as well as the other concepts introduced in this chapter have to be seen in the way Dewey and Wittgenstein conceived of concepts in general: They are instruments for further investigation. Consequently, this chapter results in a view of experts as heuristics–that is, instruments for simplifying problem solving. In particular, decision makers can choose among experts as different heuristics in planning.

7.1 TYPOLOGY OF EXPERTS

Introducing *Exploring Expertise*, Williams, Faulkner, and Fleck (1998) stressed the role of scientific uncertainty in experts' disputes:

> It would be wrong to conclude that the role of experts in controversies is solely a function of power politics. What is crucial to understand is that disagreements between experts generally take place in a context of scientific un-

certainty, with often limited empirical information and inadequate theoretical models. (p. 4)

Thus, we start with a closer look at dealing with uncertainty, too.

Formal Expertise

From the point of view of rational decision making, uncertainty has to be defined in contrast to certainty and risk: If we assume that an action is to be taken, then we can distinguish among three types of decision-making situations (Luce & Raiffa, 1957):

(a) *certainty* if each action is known to invariably lead to a specific outcome;

(b) *risk* if each action leads to specific outcomes, each outcome having a known probability; and

(c) *uncertainty* if each action leads to specific outcomes, but probabilities of the outcomes are unknown.

Usually probabilities are not known. Lotteries and certain games define situations of risks in sense (b). Most other situations, such as undergoing surgery or political decision making, are, if at all, situations of uncertainty where only the possible outcomes are known. As our discussion of financial markets and climate change demonstrated, uncertainty may even stretch further than definition (c). The development of financial markets continuously leads to new financial instruments and new regulations that redefine market conditions. Similarly, in looking for climate change policies, we do not really know all the action alternatives: a tax on CO_2 emissions, investments in clean housing and production technologies in the developing world, and so on. Thus, facing uncertainty, we can also be in a situation of incomplete knowledge about action alternatives and outcomes (see Scholz, 1983).

Models of rational decision making provide tools that capture uncertainties and structure the decision-making process. Otway and von Winterfeldt (1992), researchers on rational decision making, developed procedural rules to integrate expert judgments with the decision-making process. We can distinguish two approaches to the use of expert judgments: (a) expert judgment is simply part of the decision base, and (b) expert judgment structures and guides actual decision making. In case (a), the actual decision does not need to follow expert advice; in case (b), the whole decision process is accompanied by expert advice. Otway and von Winterfeldt called case (a) expert use "in the small" and case (b) expert use "in the large." Experts in case (b) are expert in decision making and provide formal knowledge. Otway and von Winterfeldt—formal experts as well—strongly recommended a formal expert judgment process.

Formal expert judgment process

1. Identify and select the events and options (usually called "issues") about which fact or value judgments should be made formally.
2. Identify and select the experts who will make the judgments. Strive for diversity of opinion and approaches among the experts chosen and independence in the knowledge they contribute.
3. Clearly define the issues for which judgments are elicited. This may include the decomposition of issues into subissues to facilitate making judgments.
4. Train the experts, whose expertise is in a content area, in the methodology of making formal judgments (i.e., elicitation methods, decomposition approaches, bias identification, and debiasing techniques).
5. Elicit the expert judgments in interviews with a trained elicitator, who poses questions to properly decompose the problem, to elicit the judgments, and to cross-check results against other forms of judgment.
6. Analyze and aggregate results obtained from individual experts, and, in the case of substantial disagreements, attempt to resolve differences.
7. Completely document results, including the reasoning given by the experts to support their judgments (Otway & von Winterfeldt, 1992, pp. 84–85).

Otway and von Winterfeldt's program is governed by two heuristics—namely, analytical decomposition and avoidance of judgmental bias:

(a) decompose the goal (decision aid through experts) into subgoals (using several experts), and
(b) avoid cognitive bias (such as misconceived probabilities) through formal training and analysis.

Interestingly, Otway and von Winterfeldt would have difficulties in justifying the advice to "strive for diversity of opinion and approaches among the experts" from the point of view of rational decision making alone. From this point of view, the advice would be to use the best experts—that is, experts whose judgments are most accurate and reliable. Instead, Otway and von Winterfeldt recommended some kind of expert portfolio in the Markowitz sense.

This program of formal expert judgment process has famous predecessors. After World War II, Herman Kahn propelled formal expert advice. He was project leader at the RAND corporation (for *R*esearch *and D*evelopment) that originated in a joint project of the U.S. Air Force and Douglas Aircraft. In the 1950s and 1960s, Kahn and RAND developed strategies for political planning and civil defense, such as the "credible first strike capability." In his book, *On Thermonuclear War*, Kahn (1960) emphasized a general quantitative approach:

> The major quality that distinguishes this book from most of the other works in this field is the adoption of the System Analysis point of view—the use of quantitative analysis where possible, and the setting up of a clear line of demarcation showing where quantitative analysis was not found relevant in whole or part.

Today, formal advice is provided by the technology of decision aiding (Brown, 1989), especially for governmental decision making in cases such as permitting Arctic oil and gas exploration activities or the Iran policy of the U.S. government (Brown, 1989; Brown, Laricher, Flanders, & Andreyeva, 1995). The rise of environmental politics also gave an impetus to professional decision advice on environmental risks, based on a "set of techniques from the social sciences, including system analysis, cost-benefit analysis, multiobjective optimization, and risk analysis" (Dietz & Rycroft, 1987, p. 5). Decision aid is not a modern expert service. Astrologers and theologians have also provided formal knowledge to aid decision making. Wallenstein (1583–1634), a brilliant general of the Thirty Years' War, who also showed talent both as a manager and as a political leader sought the advice of the Italian astrologer Seni—who, however, could not prevent Wallenstein from being assassinated.

A Typology of Experts

Now we are in a position to recapitulate the sorts of experts who might be addressed by "The expert"-interaction. Figure 7.1 displays the different ways for a dispositional attribution of "The expert"-interaction (see Fig. 3.3). *Dispositional attribution* means that the attribution "expert" presupposes expertise in the form of objective personal knowledge. Objective knowledge means knowledge that is not relative to a specific constellation, such as in the case of the law student who is a law expert to her friends who study humanities. Objective personal knowledge may be based on experience, as in the case of system experts, or on academic education, as in the case of today's scientists. The dispositional attribution identifies experts with their knowledge or, respectively, their expertise. Similarly, Williams, Faulkner, and Fleck (1998) concluded their introduction to *Exploring Expertise* that the role of the expert remains "socially contingent":

> A clear conclusion from all this is that the role of the expert remains socially contingent: what is judged is not so much the content of the evidence or advice, as the credibility and/or legitimacy of the person giving that evidence or advice; if we trust the expert, we must trust their expertise. (p. 4)

The typology of experts in Fig. 7.1 distinguishes between domain-specific knowledge and formal knowledge as in the formal expert judgment processes. Formal knowledge is general knowledge insofar as it can be applied to several fields; for example, mathematical knowledge and decision theory are sources of formal knowledge. We find several types of domain-specific knowledge:

- local knowledge (e.g., the different parts of Brooklyn);
- exclusive knowledge: knowledge that must be attributed exclusively to a single person or a specific group (e.g., Einstein on relativity theory in the 1920s);
- scientific knowledge (e.g., physics); and
- practical knowledge (e.g., engineering, cooking).

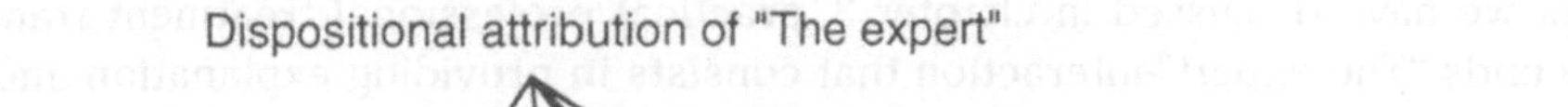

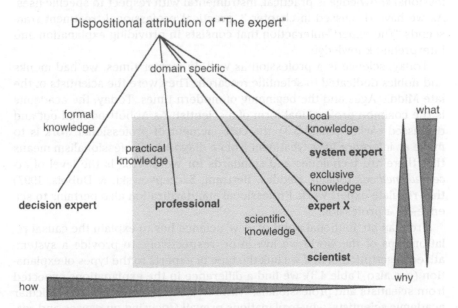

FIG. 7.1. Dispositional attribution of the social form "The expert" to a person (the expert). The expert is attributed a specific expertise: having specific knowledge and being able to explain a matter through the application of this knowledge.

The types of experts are defined according to their kind of (attributed) knowledge. Figure 7.1 also introduces the category *Expert X*. In contrast to experts in general, Expert X is a single known person who provides exclusive knowledge. Expert X appears when an expert attribution cannot be standardized: In this case, there is only this single person who has specific knowledge or specific capabilities. There are many Expert Xs in politics, sometimes leading scientists, sometimes even without any professional or academic background. Presumably there would be no use of Expert X in the formal expert judgment process because there are no standard criteria to integrate or evaluate their knowledge.

The categories in Fig. 7.1 are not exclusive to one another: A professional may act as an Expert X who is a personal consultant to a political party; a scientist may act as decision expert like Herman Kahn. In general, every domain expert can also act as a decision expert. In particular, we find nonscientific decision experts such as priests.

Let us take a closer look at the differentiation between professionals and scientists. Scientists, so to say, are the pure form of knowledge embodied in persons, whereas professionals are concerned with the practical application of knowledge. Scientific knowledge is, so to say, knowledge per se; professional knowledge is practical, instrumental with respect to specific uses. As we have discussed in chapter 3, practical professional treatment transcends "The expert"-interaction that consists in providing explanation and interpreting knowledge.

Today, science is a profession as well. In former times, we had monks and nobles dedicated to scientific research. They were the scientists of the late Middle Ages and the beginning of modern times. Today, the academic is the common professional form of a scientist. As Abbott pointed out and discussed earlier (chap. 3.3), the core element of professional work is to make an inference for a treatment from a diagnosis. Professionalism means that there are techniques and standards for work. This is the level of *accepted methods* (Shalin, Geddes, Bertram, Szczepkowski, & DuBois, 1997) that regulate expert work. Professional standardization also pertains to science—as a profession.

From an attributional point of view, science has to explain the causal relationships of the world we live in or, respectively, to provide a systematized description of it. If we link the type of experts to the types of explanation (see also Table 4.3), we find a difference in the explanations expected from scientists and professionals: With reference to scientists, in particular academic scientists, why-explanations prevail (focusing on causes and reasons), whereas expectations from professionals relate more to procedures (= how-explanations). About the nature and functions of viruses, we would like to ask scientists, but we would probably prefer to ask the doctor how to take precautions when traveling.

To complete the picture: System experts have to provide what-explanations; they know the state and history of a system—a town or geological site. Decision experts need not know much about the content of a problem; they provide how-explanations and procedures. The type of explanation by Expert Xs depends on what kind of advice and exclusive knowledge they provide in particular cases. Whatever the type of explanation might be, all experts act as interpreters of certain knowledge that is attributed to them by nonexperts.

7.2 UNCERTAINTY ≠ INSECURITY

Margolis' Risk Matrix

In 1996, Howard Margolis of the University of Chicago published a book, *Dealing with Risk: Why the Public and the Experts Disagree on Environmental Issues*. He argued that:

1. Human judgment under uncertainty has a "propensity to dichotomous responses." For instance, people have a bimodal response in their willingness to obtain insurance for small risks. Either the person treats the risk as negligible and is not willing to pay much for insurance: Or the person treats the risk as significant and is willing to pay much more than the expected value of the risk (see McClelland, Schulze, & Coursey, 1993).

2. As to environmental issues, controversy results from the differences in accepting a trade-off between costs and benefits. Margolis explicates the point using a dichotomous risk matrix (see Table 7.1). The risk matrix includes two dimensions: danger and opportunity. On the one hand, the risk matrix includes the danger that arises if no political or societal precautions are taken—for instance, from nuclear power or from asbestos contamination. On the other hand, it includes the benefits of certain technologies (e.g., nuclear power) or, respectively, the costs of taking precautions—for instance, for removing all asbestos.

TABLE 7.1
Risk Matrix According to Margolis

		Opportunity	
		Yes	*No*
Danger	Yes	Fungibility	"Better safe than sorry"
	No	"Waste not, want not"	Indifference

Note. From *Dealing with Risk* (p. 76) by H. Margolis, 1996, Chicago: The University of Chicago Press. Copyright 1996 by The University of Chicago Press. Reprinted by permission.

There are two dichotomous reactions toward environmental risks. Margolis named them with proverbs. The first is "better safe than sorry," which signals caution and means that an issue is taken seriously. The second is "waste not, want not," which signals "getting on with things" (p. 74). There is also a third position that Margolis called *fungibility*; in this position, we trade off benefits and costs: "We are in a situation where we can see a need to consider, and usually a need to somehow balance or trade off, the advantages of caution against the advantages of boldness" (loc. cit., pp. 76–77).

Controversy on an environmental issue starts when the issue is interpreted in a dichotomous way by experts and nonexperts. This is apparently more often the case with subtle and small "new risks" that are difficult to interpret. Margolis pointed to what he called a paradox that "characteristically very small risks come to be the focus of salient lay concern as a loss from what had been taken as the normal situation" (p. 96). Margolis added that such controversies between experts and nonexperts might be a form of the general social and political polarization: "I would argue that these polarized expert/lay cases for environmental risks should ultimately be seen as a special category of the more general phenomenon of social and political polarization" (loc. cit., p. 92).

3. According to Margolis, experts in general trade off danger and costs for precautions or, respectively, costs and benefits of technologies. This is due to the simple fact that "an expert by definition has a lot of experience in an issue where a lay person by definition is on unfamiliar ground" (p. 95).

> Expert/lay conflicts can arise for both situations—individual or social choice—because in any context at all, there will be cues that are subtle or complicated or otherwise difficult to use for an inexperienced person, but are familiar and automatically significant for the experienced person. Just what we mean by *expert* is that a person has a lot of experience on some matter. (loc. cit., p. 79)

Unfortunately, there is a flaw in Margolis' argumentation. There are environmental issues where the controversy is exactly the opposite. In some cases, ecological experts are on the "better safe than sorry" side and the public on the other side. This is true for the discussion on biodiversity. To preserve biodiversity—saving "unpopular" forms of animals and plants—is an almost academic concern that does not allow for trade-offs. A similar case was the experts' concern about the potential of natural resources—such as coal or chrome—expressed by the report to the Club of Rome on *The Limits to Growth* (Meadows, Meadows, Randers, & Behrens, 1972).

Margolis' argument operates with a weak notion of *expert*. He defined the *expert* as an experienced person. Yet the controversy he wanted to explain is between a class of scientists-engineers and parts of the public. However these experts might be experienced in their particular field of expertise, they cannot have—by definition—any experience with a small new risk, such as in some nuclear issues for the simple reason that the risk is too unlikely to become subject to the experience of experts. As to these forms of risks, only a weak-form expertise is possible. The expertise of physicists and engineers in nuclear technology is not based on experience, but on abstraction in the sense that Abbott introduced: By virtue of their professional-scientific knowledge they claim to have competence on nuclear issues, too. Thus, the controversy is not on fungibility, but on the kind of expertise: Parts of the public do not trust scientific experts to really have sufficient knowledge on small new risks. They refuse the scientific explanations that link new phenomena, such as nuclear power or BSE, to general laws or state-of-the-art scientific knowledge.

Using Experts

The dichotomies Margolis observed are not based on differences in knowledge—the experts know about fungibility, the public does not—but on differences in concern: The public feels concerned, whereas the experts do not. To understand public concern, we introduce a distinction between uncertainty and insecurity. *Uncertainty* means lack of knowledge. Uncertainty is an epistemic category and linked to the question, "What do we know?" In contrast, *insecurity* means need for control. Insecurity is a pragmatic category and linked to the question, "What shall we do?" Uncertainty and insecurity are not always correlated. Most uncertainties that interest scientists such as mathematicians or philologists do not concern the public. Of course, there are fields where both uncertainty and insecurity are high, notoriously in the case of politics, although some political scientists or philosophers might claim to be able to predict political developments with certainty. In the fields of less insecurity but high uncertainty, we find the more private beliefs and interests invested in hobbies or religious practices.

The distinction between uncertainty and insecurity is a means to characterize different uses of experts or expertise, respectively. For this purpose, we define a field to be of *low uncertainty* if there is sufficient formalized knowledge—for instance, in most engineering problems or typical health problems. We define a field to be of *high insecurity* if there is discussion on precautions that need to be taken—for instance, in foreign policies or, again, in some typical health problems. Table 7.2 shows several combinations of insecurity and uncertainty. The fields differ in the kind of dominant expert services. There are also fields where society accepts insecurity as relatively high but sees uncertainty as relatively low. This is the field of common professionalized problems such as health care, judicial procedures, or accounting, for which problem solutions are offered by professionals.

TABLE 7.2
Type of Services and Type of Experts That Prevail Under Special Conditions of Uncertainty and Insecurity (as Perceived by Society)

		Insecurity	
		++	+
Uncertainty	++	Consulting through Expert X [political agendas]	Support through system experts [beliefs & interests]
	+	Treatment through professions [problem solutions]	Discussion through scientists [scientific paradigms]

Note. In brackets are the type of knowledge field where the expert services compete. The cells are not exclusive to one another; for instance, we find scientists also doing many other types of service.

We distinguish among consulting, treatment, discussion, and support. In the case of professional services, we have expert treatment, such as surgery or advocacy. Like Abbott (1988), we consider professional treatment as a form of problem solving. If a problem of some social importance is both of low uncertainty and insecurity, it may only receive attention in scientific circles. This is the case in ancient philology, the study of ancient languages such as Latin or Sanskrit. We can term the type of expert work provided by scientific experts as *discussion*. Scientific discussion takes place in the form of journal science (Fleck): Pieces of research are used as arguments to defend or attack scientific theories. The field of competition is defined by paradigms, as described by Thomas S. Kuhn (1962)—that is, scientific schools or research groups that have established a specific way of theorizing. For instance, psychoanalysis may be considered a paradigm among all the different schools of psychotherapy.

In the case of high uncertainty and high insecurity, we find consulting through a personal Expert X who can also refer to a professional association or other prominent companies such as Greenpeace or McKinsey. Consulting by Expert Xs takes place in corporations and politics. We can consider the International Panel on Climate Change (IPCC) as an Expert X on the science of climate change for the international community; we have already discussed the function of scenarios that transform scientific knowledge into policy advice.

The least typical kind of expert service is support, particularly support through system experts. This kind of expert service is often free of charge: System experts are considered as incorporating the system knowledge, thus, information given by system experts is some kind of a favor or courtesy. We find support through system experts, for instance, when we ask a farmer about the local weather and soil conditions, or the president of a golf club on how to apply for membership. We have already discussed the use of system experts with regard to environmental problem solving.

Table 7.2 does not comprise relative experts–that is, the experts by occasion who give a piece of information (where to get to the station?) or a piece of advice (how to tie laces?) for the reason that there is normally no field of competition between such occasional expert services. It should also be noted that the use of experts might base on "The expert"-interaction, but in general transcends it–as we have already discussed for the case of professional treatment.

Table 7.2 shows the relationships among uncertainty, insecurity, and expert services. If we want to put a scientific subject onto the political agenda, it is necessary to increase public concern–that is, we have to increase perceived insecurity. Furthermore, if we want to maintain a scientific outlook, we have to promise to reduce uncertainty, for instance, by systematizing the problem aspects or inventing formal problem-solving procedures.

The distinction in Table 7.2 parallels the finding by Donald MacKenzie (1998)–what he called the *certainty trough*. He argued that the uncertainty that any particular technology causes varies according to one's viewpoint. If we change from the point of view of the developer to that of the user, this uncertainty sinks. However, it rises when we change to the viewpoint of the developer of a different technology in the same filed of application. This means that we have a shift in uncertainty perception, when plotted resulting in a trough shaped graph with unequally high sides. On the lower side, we have uncertainty perceived by the developers of the given technology; on the other higher side, we have uncertainty perceived by developers of a different, but related technology. The bottom of the trough is characterized by uncertainty perceived by users of the given technology. Their perception of uncertainty is therefore smallest, followed by the developer's perception. Uncertainty is perceived highest by developers of other, but related technologies.

Therefore, MacKenzie called this phenomenon *uncertainty trough*. When interpreting the uncertainty trough from the perspective of Table 7.2, we can say: One scientific paradigm might consider a different one as "not serious," thus perceived uncertainty increases. However, professional work proceeds as if this technology were an accepted method, thus, perceived uncertainty decreases.

7.3 EXPERTS AS HEURISTICS

We turn back to experts as interpreters of certain knowledge that is attributed to them by nonexperts. This section focuses on expertise connected to human–environment systems, also including marketlike environments. The case studies in chapters 5 and 6 were both concerned with the predictive potential of knowledge about human–environment systems, the examples being climate change and financial markets. We now discuss the role of knowledge and expertise in prediction. Our starting point, once again, is risk controversies.

Compensatory Versus Noncompensatory Models

We might suppose that risk as variance is merely a scientific risk assessment, whereas risk as commitment is the reality of risk communication. However, this is too simple a view. Instead, we return to Margolis' risk matrix (Table 7.1) and try to determine why, given a critical environmental issue, some persons assume "fungibility" (the scientist?) and others do not (the public?). We precluded Margolis' explanation that scientists, because of their experience, know even the small new risks and therefore can trade off the pros and cons. The point is rather that risk judgments are relative to models of nature in the sense introduced by the integrated assessment approach. For the context of our discussion, we distinguish between compensatory and noncompensatory models.

In compensatory models, we can compensate every loss with an equivalent gain. In such models, it is always possible to achieve an equilibrium. This is a central idea that guides models of financial markets. As to environmental issues, the benign model and the perverse/tolerant model of nature seem to some extent compensatory (see Fig. 6.5). In noncompensatory models, some losses cannot be compensated. Moral systems tend to be noncompensatory because of absolute norms that may not be violated in any case (Mieg, 1991)—for instance, the unsurpassable value of a single human life. Ecology, too, argues with noncompensatory models such as specific ecosystems. From the point of view of the integrated assessment approach, noncompensatory environmental models are ephemeral models of nature (see Fig. 6.5).

The distinction between compensatory and noncompensatory models supports Margolis central idea of fungibility. In the case of compensatory models, fungibility is possible: We can trade off cost and caution, gains and losses. Table 7.3 depicts further characteristics that correlate to the difference between compensatory and noncompensatory models. For instance, from an ethical point of view, compensatory models allow for utilitaristic interpretation. We should act in a way that optimizes the utility for all who are concerned by our actions—or, to put it in the words of the philosopher Bentham, we have to strive for the "greatest possible quantity of happiness." In contrast, the ethical interpretation of noncompensatory models would be categorical, stating that there are strict limits of utility. For instance, in his biocentric ethics, Taylor (1986) claimed that each form of life has an intrinsic untouchable value, thus claiming "respect for nature."

TABLE 7.3
Compensatory Versus Noncompensatory Models
for Complex Human–Environment Systems

	Compensatory Models	*Noncompensatory Models*
Examples	Nuclear power (Margolis, 1996) Stock markets	Ecosystems (Pahl-Wostl, 1995) Democratic constitution
Experts' risk communication (in terms of Margolis, 1996)	Fungibility	"Better safe than sorry"
Ethical interpretation	Utilitarism?	Categorical?
Prevailing risk interpretation	Variance? (as-if-we-all-knew clause)	Commitment? (Time-&-commitment clause)
Predictability	Limited by systemic effects?	Limited by uncertainty?
Heuristics	Forward? (Diversification, securization, etc.*)	Backward? (Preservation, reducing contamination, etc.)

Note. The distinction between these two types of models is hypothetical. There might be many systems or data that do not fit into this scheme. Therefore, we use question marks.

*Securization means preferring tradeable investments (e.g., a share) to untradeable ones (e.g., a loan).

With some reservation, we can class the two risk interpretations that we introduced. Provided that we have an adequate measure of outcomes (such as money), compensatory models allow us to regard risk as variance. This is the case in financial markets, where a gain can directly compensate a loss. In noncompensatory models, decision making is committed to the values at risk: maybe climate or constitution. However, we do not want to overstress the distinction between risk as variance or risk as commitment: We could even describe the behavior of investors in terms of risk as commitment, particularly in cases where no information or evidence really supports a decision to invest in a specific security.

Predictability, Heuristics, Planning

The points we want to discuss in detail concern predictability and heuristics. As to climate change, we found predictability limited by many uncertainties: uncertainty about the impact of some elements of the climate system (such as clouds, aerosols, etc.), uncertainty about central parameters (such as climate sensitivity), and the so-called *social and economic uncertainties.* We do not really know whether the climate system is noncompensatory. There are, of course, compensatory factors such as the oceans that absorb CO_2. We lack knowledge of alternatives to the present state of the climate system. Perhaps there are other states of equilibrium as claimed by a tolerant perspective. In financial markets, there are uncertainties, too. As discussed, the limits of predictability are presumably set by systemic effects—that is, feedback effects of attempts to regulate the systems.

One of the most influential psychological papers on predication was published by Kahneman and Tversky in 1973. They claimed that human predictions tend to be predominated by representative features and, in that case, neglect probabilities. Kahneman and Tversky conducted an experiment: They made subjects predict the field of studies of students for whom they were provided personality descriptions. Subjects also were given the probabilities for the fields of study (more correctly speaking: the frequencies of students in various fields). Kahneman and Tversky found that subjects often judged according to similarities between the personality descriptions and certain fields of studies ("a typical law student"), even when the probability for the field of study was low. We do not discuss the general question of how appropriate the Kahneman and Tversky model is for human prediction (see Gigerenzer & Murray, 1987; Mieg, 1993).

However, as we saw, prediction is confounded with planning. We discussed the case of scenario formation: We can model uncertainty by depicting a variety of scenarios such as the IPCC scenarios on climate change. The problem is, even if we can say with certainty what will happen once a scenario becomes reality, it is difficult to determine the probability of the occurrence of the scenarios.

For instance, the IPCC 1990 Scenario D predicts that CO_2 emissions will be reduced to 50% of 1985 level through "accelerated policies": But what is the probability for realizing "accelerated policies"? This is clearly a question of politics or policy planning. In addition, however probable scenarios are, they have to be consistent. For instance, "accelerated policies" would be inconsistent with allowing for unlimited use of fossil fuels. We can say: Internal consistency is a necessary condition for scenarios, probability is not. Thus, in cases where prediction and planning are confounded, we have to look for similarities to obtain consistency. If we had to award someone a grant for studying law, we would be well advised to award it to someone whose personality description fits a law student. Thus, in such cases, contrary to Kahneman and Tversky (1973), it would be rational to judge or predict according to representative features to increase consistency.

Kahneman and Tversky (1973) called the cognitive tendency toward representativeness a *heuristic*, a rule of thumb. In general, Kahneman and Tversky depicted heuristics as cognitive bias (especially Kahneman, Slovic, & Tversky, 1982), whereas Herbert A. Simon introduced heuristics as some kind of simplified problem-solving procedure. For instance, according to Simon, the planning heuristic simplifies problems by "omitting some of the detail" and "abstracting essential features." We may think of scenario construction as an instance of the planning heuristic emphasizing certain distinctive paths to the future. Table 7.3 also displays some instances of heuristics for explanation and problem solving. The obvious difference between heuristics for compensatory models (such as financial markets) and noncompensatory models (such as ecosystems) is that the heuristics for noncompensatory models try to reconstitute the presumed normal state—for instance, by reducing CO_2 emissions. They are backward-oriented to ensure preservation. Heuristics for compensatory models allow for forward strategies—for instance, diversification and portfolio selection. They are oriented toward change.

From the point of view of collective cognition, we can consider experts as heuristics, too. Experts embody specific experience (we refer to experts through dispositional attribution as in Fig. 7.1). By using an expert, we use the specific experience of someone else to solve a problem or find an explanation. We can choose among different experts: For educational problems, we can consult a teacher, a priest, or maybe our own parents. Each of them has his or her way to see and explain the problem. The result is different kinds of advice. As discussed earlier, the specific *Leistung* of an expert is to provide experience that everyone could reach provided he or she had enough time. Thus, by using experts, we decrease our own cognitive effort. We use experts as heuristics. Little wonder that Kahneman (1991) is as skeptical toward experts as to heuristics in general.

The social aspect of using experts as heuristics is: Experts are units of accountability. In principle, expert service can be tested. Thus, to some extent, the outcome of the expert service can be attributed to the expert. The test criteria depend on the kind of explanation, as displayed in Table 4.3. For instance, expert predictions in financial markets can be compared to the actual development of the market. The criterion simply asks whether the prediction is true—for instance, whether the predicted price is close to the actual price of a stock. For decision experts (e.g., for conflict mediating in a family), the criterion is effectivity: Could the conflict be settled or not? In climate change, there are no efficient selection criteria for experts. In this case, the expert status has to be attributed in accordance to professional, academic standards. Professions regulate accountability. In general, by means of licenses, control of education, and professional standards, professions such as medicine try to eliminate external competition (Baer, 1986; Freidson, 1986a). Instead, they install a moderate, regulated internal competition among their own members. As Abbott (1988) showed, there is competition among the professions, and some professions lose influence such as astrologers or even disappear completely such as the train physicians of the emerging 20th century.

The problems of using experts as heuristics can be characterized in analogy with heuristics in general (see e.g., Groner, Groner, & Bischof, 1983; Mieg, 1993): Heuristics have application conditions that are too general and no reliable solutions. In addition, problems may arise due to uncertainty of systemic effects. Three critical points as to experts have to be added:

(a) *Too general application conditions.* The conditions for applying experts to problems often remain unclear. Professions try to regulate the specification of expertise within the domain of the profession. Thus, we have various medical specialists such as radiologist, psychiatrists, dermatologist, and so on. Experience and expertise are domain-specific, but there is interprofessional competition. The application of the expertise of one domain to problems of another domain (e.g., "engineering" for problems of business administration) can be heuristic in the true sense of the word: That means it leads to new kinds of knowledge and problem solutions.

(b) *No reliable solutions.* Using experts does not always yield a problem solution. Again, professions try to standardize knowledge for routines. Today, surgery of an appendix is routine work. The same is true for short-term weather forecasts. As Shanteau (1992a, 1992b) showed, expert performance measured as accuracy of predictions depends on the domain. There are domains where experts show high performance (e.g., in meteorology or physics), and there are domain experts with poor performance, such as stockbrokers or clinical psychologists (see Table 2.3). As we have seen, specific investment strategies might be successful only for limited time.

(c) *Systemic uncertainty.* Expert service may increase systemic uncertainty. As discussed for financial markets, in some fields, we find an intricate recursivity in expert service: The decision or action of one decision maker might provoke a new type of response by other actors that would be impossible without the preceding action of the counterpart. Thus, the whole system develops. We saw the exigency of derivates as a form of insurance against losses in financial contracts that, once in existence, could be used as new instruments of speculation that require new forms of regulation. We can also presume this kind of recursivity in psychiatry, in courts of law, and in other fields where predictions by experts demonstrate only limited performance.

Using experts as heuristics also means that experts are potentially subject to selection. We saw the use of relatively blind experts in predicting financial markets. We also saw the inclusion of system experts in assessing environmental problems. Selection of experts can be subject to different criteria (see also Fig. 7.1 & Table 7.2). We may select expert according to the:

- type of knowledge (investments, climate change, etc.),
- type of service (consulting, discussion, treatment, support, etc.),

- type of expert (professional, scientific, decision expert, system expert, Expert X), and
- type and degree of *Leistung* (what-, how-, why-explanations; individual performance; etc.).

Before we discuss selection criteria, it should be noted that to understand experts as heuristics is in no way a new insight. Obviously, the idea dates back to evolutionary theory in the 19th century, particularly to the work of Charles Darwin. In 1960, Campbell wrote a paper on the "blind variation and selective retention in creative thought as in other knowledge processes," arguing that the set of creative personalities is subject to variation and selection. In the same spirit, D. K. Simonton (1988) has done research on the psychology of science and revealed, for instance, the enormous productivity of high-ranked scientists.

This book, *The Social Psychology of Expertise*, does not want to add further theorizing in this direction. Rather it aims to come to conclusions from our discussion on prediction, experts, and uncertainty. As we have seen, there is a demand for experts in financial markets as well as the climate change issue, but expert predictions have been found limited in both cases. Hogarth and Makridakis (1987)—both decision experts—described some lessons from research on forecasting and planning. Table 7.4 links these lessons to the problems of using experts as heuristics. Hogarth and Makridakis (1987) recommended:

(a) To avoid biases in information acquisition it is necessary to sample "information from as wide a base as possible." To avoid accepting false forecasts in haste one should find "disconfirming" data and hypotheses.

TABLE 7.4
Performance Related Criteria for Selecting Experts (as Heuristics) and Lessons From Forecasting and Planning (Hogarth & Makridakis, 1987)

Performance Criterion	*Problem of Heuristics (Groner, Groner, & Bischof, 1983; Mieg, 1993)*	*Lesson From Forecasting and Planning (Hogarth & Makridakis, 1987)*
Validity (Do expert judgments describe reality?)	Too general application conditions; domain-specific effectivity	Use different experts; find disconfirming data and hypotheses
Reliability (Do repeated expert judgments lead to the same results?)	No reliable solution	Information aggregation should be done mechanically
Leistung (Is the expert's service efficient and effective?)	Systemic uncertainty through conflicting heuristics	Carefully interpret causes; if possible, use simple models

(b) As people are inefficient in aggregating information, this should be done mechanically.

(c) To avoid false attributions of apparent causes ("illusion of control"), greater care needs to be exercised when interpreting the causes. (p. 548)

Hogarth and Makridakis added that, in the context of forecasting and planning, simple methods of prediction such as simple average of expert group opinion often outperformed the more sophisticated ones.

If we think of experts as heuristics and look for criteria for selecting experts or expertise, respectively, we can treat experts in the same way as tests. The two main criteria for psychological tests are validity and reliability, with *validity* roughly meaning that the test mirrors reality, and *reliability* meaning that a second equivalent measurement with the same test will lead to the same results. If we wish to have valid expert judgments, we should take into account that expertise is domain-specific. As we have seen, similar problems arise with heuristics in general: We have to specify the application conditions. If we are still not sure about the validity of the expert judgments, we can resort to Hogarth and Makridakis (1987): They advise us to use different experts (of the same domain) and look for disconfirming data. As we see, we can match validity as a selection criterion to the discussion of problems of heuristics and the lessons by Hogarth and Makridakis. This is what Table 7.4 shows. To increase reliability, the advice according to Hogarth and Makridakis would be to mechanically aggregate information from different experts or from different single judgments by one expert.

Table 7.4 adds to the two criteria of validity and reliability a third one called *Leistung*. We said: The *Leistung* of an expert (effect and efficiency) for the sake of which experts are used is the relatively fast utilization of the expert's compressed experience any reasonable person could make if he or she had enough time to do so. Thus, for the selection of experts, we should consider:

- to what extent expertise by experience seems possible in the particular field,
- to what extent can expertise in the particular field be communicated, and
- what are the—unintended—effects of using an expert?

From the perspective of experts as heuristics, we should be aware of increased systemic uncertainty caused by the use of heuristics; we might also step into problems of evaluating and singling out the effects of heuristics. A lesson from research on forecasting and planning would be to carefully interpret cause and use simple models if possible.

Table 7.4 shows the criteria for expert services from a psychological point of view. Nevertheless, selecting experts is a managerial task. To come to more detailed conclusions on how to use experts, we would have to develop a systemic view of knowledge management that would be within the scope of cognitive economics. This is outside the scope of this text. Some conclusions on an expert role approach in management are drawn in chapter 8.

CHAPTER

8

Conclusions for Management With Experts: The Expert Role Approach

To conclude, we discuss the role of experts in managerial decision making. Managers are confronted with experts in several forms, mainly consultants and specialists as members of their staff. The aim of these conclusions is not to recommend additional advising experts. This would mean overloading managerial decision making. For the case when management involves experts' advice–such as any kind of consultancy–this chapter shows how to use *The Social Psychology of Expertise* to use experts and expertise more efficiently. The essence of this chapter 8 is the introduction of an expert role approach that shall support robust action in management. Chapter 8 commences with some fundamentals of management theory and then introduces the expert role approach. This chapter–and the book–ends with some remarks as to this question: When do we need experts for planning?

8.1 PARADOXES OF MANAGEMENT THEORY

Makridakis (1990), a doyen in the field of management theory, described the paradoxes of using management theories:

1. A manager must use a management theory to guide his/her thinking and facilitate or improve his/her decision-making, yet past experience indicates that the great majority of theories in management are short-lived and of doubtful value [. . .]
2. A manager needs a predictive theory, yet in management few, if any, theories are predictive.

3. A manager needs a simple theory which he/she can easily understand and which can tell him/her how to succeed, yet no such theories can be found in the public domain. (p. 20)

Let us consider *theory* as a general word for cognitive decision aids such as a concept, plan and strategy, or heuristic (i.e., a rule of thumb). Having discussed expertise and experts in various contexts, we can understand the paradoxes of management as described by Makridakis as critical aspects of management as expertise and of management with experts. Chapter 5 showed such a case in the context of financial markets: We need financial experts who can predict the market, but expert predictions have been found to be rather limited in this domain. In general, problems for management with experts arise from the fact that:

- management in general concerns human behavior,
- experts should provide information, and
- there are organizational constraints for the use of experts.

Let us examine each point in detail.

Management Is Concerned With Human Behavior

Discussing Shanteau's list of domains of expertise (see Table 2.3), we saw that expertise is notoriously poor for processes that involve human behavior. This is true for personnel selectors and clinical psychologists, and we might suspect that it is true for management as well. If we try to predict and control human behavior, we face the problem of the autonomy of the persons involved. Any theory or expectation concerning the behavior of a person can be the reason for this particular person to act in a different way than we predict or expect. In his book *The Alchemy of Finance*, Soros spoke of the reflexivity of social actors that also causes the relative unpredictability of financial markets. In a similar way, we discussed the behavior of financial markets in terms of recursivity of financial market functions: One actor decides on the basis of the expected or factual decisions of others. Thus, every regulation for risk management can be used for taking over even riskier investments, hedging being the most prominent example.

The same is true in most decisions on behavior. Usually a client knows what the psychologist or personnel selector expects him or her to do and can behave according to this expectation. From a general point of view, the problem of reflexivity distinguishes social sciences from natural sciences. In natural sciences, the scientific object does not usually react to the specific theory or expectation the scientist has about it.

The questionable utility of expert assessment in management has been shown by Clayman (1987). She analyzed the fate of U.S. firms between 1980 and 1987 after Peters and Waterman's (1982) *In Search of Excellence* described them to be innovative and financially successful. According to Peters and Waterman, excellent firms are "brilliant on the basics" (p. 13); in particular, they are characterized by "productivity through people"—that is, they "treat the rank and file as the root source of quality and productivity gain" (p. 14). Clayman compared these firms with the group of firms within the S & P 500 population that had performed worst according to the financial criteria from Peters and Waterman (1982), such as asset growth and return on equity. For the "excellent firms," the rate of asset growth value nearly halved after 1980. Investments made in 1980 in stocks of "unexcellent" firms tripled in value by 1987, clearly outperforming investments in stocks of excellent firms that grew to about 180% by 1987.

Experts Should Provide Information

We said that experts are knowledge interpreters (chap. 3). An alternative to knowledge interpretation could focus on knowledge as such. Managers who do not wish to depend on experts have to consult books and other information-storage systems. As we have already discussed, data are not knowledge. Data need interpretation. There are data (e.g., information on prices) that do not need much interpretation. Other data (e.g., on soil contamination) do need interpretation. This difference is quite evident when we consider which kind of industry uses relatively uninterpreted data. Table 8.1 shows the results of a study by the European Commission on the markets for electronic information services in the European economic area (European Commission, 1996). The study reveals the predominant position of financial services in using electronic information services—although the financial sector is far from being the most productive industry. We can as-

TABLE 8.1
Revenues of the European Information Industry
by User Groups in 1994 in % of Total Industry Revenues

User Group	*Revenues of the Information Industry**	*Gross Value Added (GVA) of User Group***
Financial services	41.3%	287'949 (5.0%)
Other business/corporate	38.8%	1'213'667 (21.0%)
Government/public sector	10.9%	536'722 (9.3%)

Note. The use of information systems prevails in the financial sector, which is not the dominant industry in terms of GVA (gross value added). We can assume that—at least for managers—financial data need less interpretation than data in other industries.

*Source: European Commission (1996, Table 16).

**GVA in millions of ECU at current prices. Source: Eurostat (Nace-Clio), grouping: "credit and insurance institutions," "manufactured products," "general government services."

sume that—at least for managers—financial data need less interpretation than data in other industries.

Organizational Constraints for the Use of Experts

We consider an organization to be a corporate actor in the form of a resource pool (see chap. 4): Individuals allocate parts of their resources (time, capabilities, money, etc.) to a central entity (the organization). This involves the simple but important distinction between resources that are part of the particular organization and resources that are not. Using experts and expert knowledge can put the distinction between internal and external resources into question. There is a dilemma between relevance and autonomy of expert knowledge: On the one hand, if we use external sources of expert knowledge (e.g., business consultants), information might be irrelevant and/or not specific enough for our company. On the other hand, if a company totally integrates experts, they are likely to lose both contact with science and their capability of autonomously developing their knowledge base:

> This happens to large insurance companies such as Swiss Re that employ specialists in natural sciences. Usually, the companies train these specialists in the insurance business. As a result, the specialists become underwriters and lose their scientific competence. Therefore, these companies have to find other organizational forms of combining the companies' need for useful scientific expertise and the experts' need for developing their knowledge. The Swiss Re created a sort of think tank for risk assessment with scientists who conduct their own projects but also serve an internal expert service function.

So far, research on experts in organizations has focused on role conflicts between professional autonomy and bureaucratic administration. In chapter 4, we discussed the study by Hall (1968), who found a negative correlation between criteria for bureaucratization and the perception of autonomy. However, 30 years of research on the role conflicts of professionals encountered in bureaucratic organizations have not yielded consistent results. We argued in chapter 4 that an analysis of the expert's role reveals the organizational role conflicts with experts as well as possible resolutions. In the following paragraph, we compile the elements of an expert role approach and apply it to management.

Standard Solutions: Commissions and Consultants

Many companies provide specific positions for experts (e.g., in law or market research divisions). In such a way, bureaucratic administrations link types of problems to types of specialists. If there are nonroutine tasks or if external expertise is required, the standard solutions are to install a commission or use a consultant.

In general, business consultants provide support for managers on the basis of a professional contract. According to James O. McKinsey, being a consultant requires as much "professional exposure" and "unquestioned respectability" as "a reputation for expertise in an area of some concern to management" (cited in Neukom, 1975, p. 3). In general, the problems arising with consultants are exactly the problems discussed earlier—namely, organizational relevance of the consultant's advice and questionable reliability of predictions.

In contrast to consulting, the organizational solution of installing a commission can overcome or reduce the problems mentioned previously by the selection of the participants (e.g., in cross-functional teams used in business process reengineering; Davenport, 1993; Hammer & Champy, 1993). Yet commissions suffer from well-known disabilities. A commission:

- takes time,
- costs (in particular, lost working hours of participating internal personnel),
- risks ending without results,
- risks resulting in installing new subcommissions, or
- risks proposing solutions that cannot be implemented.

From the point of view of social psychology, commissions are temporary groups. Judgment and problem solving in groups, on the one hand, show certain motivational and cognitive distortions, such as individual loafing, compliance/conformity, and risky shifts (for an overview, see Levine, Resnick, & Higgins, 1993). On the other hand, groups can—but not necessarily do—outperform individuals when it comes to problem solving. As discussed in chapter 4, in this case, groups assign expert roles for individuals with particular expertise within that group.

8.2 EXPERT ROLES AND MANAGEMENT

The Social Psychology of Expertise has implications for theory and practice of managerial decision making. The implications constitute what we can call the *expert role approach*. Table 8.2 is an overview of the main concepts and findings related to this approach that are assembled in this book.

Professional Work

The basic assumption of the expert role approach consists of the link between types of experts and the kind of service experts can provide. We started with Abbott's description of professional work, which we already

TABLE 8.2
Overview of the Conceptual Elements of the Expert Role Approach Displayed in Previous Chapters

Concept		*Where to Find?*
Types of experts	General view: "The expert" is an attributed social form of interaction.	Figure 3.3
	Types of experts in terms of attributed competence.	Figure 7.1
	Types of experts with reference to the types of their work.	Tables 7.2 & 8.3
The expert's role	General view: The attribution of "The expert" depends on the specific social situation, in particular on the kind of problem to be solved; the kind of knowledge that the expert provides; and the kind of audience or clients in that situation.	Figure 4.2
	Kinds of legitimizing uses of experts.	Table 4.1
	Role conflicts with experts and their resolution.	Table 4.2
Expert performance	General view: In some specific domains, we consistently find high expert performance, in others not; we find particularly poor expertise when judgments about human behavior are involved (Shanteau's list).	Table 2.3
	Lessons from forecasting and planning for the use of experts (Hogarth & Makridakis).	Table 7.4
	Professional work (Abbott).	Figure 8.1

showed parallels concepts and findings in cognitive psychology (chap. 3.3). Abbott (1988) described the following sequence of professional work: diagnosis, inference, and treatment (although he did not use the same sequence in describing them):

- *Diagnosis*. "[D]iagnosis first assembles clients' relevant needs into a picture and then places this picture in the proper diagnostic category" (p. 41).
- *Inference*. Inference is a "purely professional act" (p. 40). "It takes the information of diagnosis and indicates a range of treatments with their predicted outcomes" (p. 40).
- *Treatment*. "The effects of treatment parallel those of diagnosis. Like diagnosis, treatment imposes a subjective structure on the problems with which a profession works" (p. 44).

Figure 8.1 gives an impression of the sequence of professional work.

For the concern of managerial decision making, we distinguish four types of experts: professionals, decision experts, scientists, and relational experts. By *relational experts*, we refer to system experts and all persons

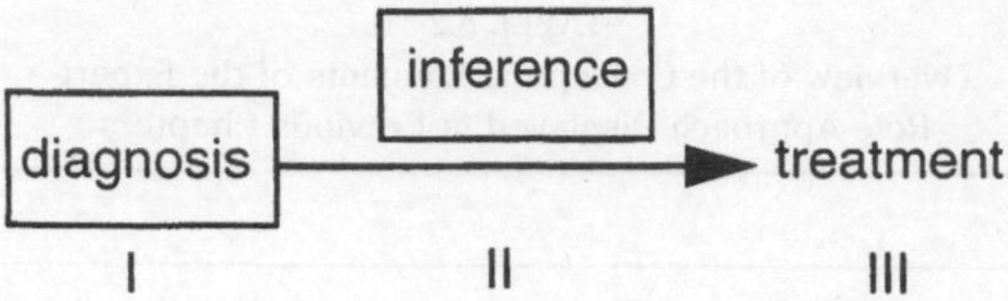

FIG. 8.1. Sequence of professional work (adapted from Abbott, 1988).

TABLE 8.3
The Expert Role Approach: Link Between Type of Experts and Type of Services

Type of Expert	*Type of Service (Also Scope of Accountability)*
Professionals	Professional task (Fig. 8.1)
Decision experts	Inference
Scientists	Diagnosis (analysis)
Relational experts	Information

who are addressed as experts in specific circumstances, but not because they are scientists or professionals. For example, in every organization, there is always one person who—without any authority—seems "to know everything and everyone" and whom we can ask regarding the organization's informal structure. Relational experts can also be experts by role assignment (e.g., in teams where each member is responsible for a specific part of the team's work).

Table 8.3 shows the links between types of experts and types of services. In contrast to professionals, scientists are concerned with one component of professional work—diagnosis or analysis. Decision experts focus on the inference component. Relational experts provide information and minimal, data-oriented explanations. An expert's role also determines the scope of accountability of the expert's work—as discussed in chapter 3. Professionals account for the whole professional task, including treatment. Scientists and decision experts account for their analyses. Relational experts only account for the information they provide. This particular link between a kind of service and accountability is—to put it another way—the expert's responsibility (see Mieg, 1994).

Identifying Experts

Stein (1992, 1997) developed a method of identifying human experts in organizational contexts. He promoted the view that "considers the expert in relation to a referent social group such as a social network, an organization, a constituency, or a market" (Stein, 1997, p. 182). Expertise is a social variable. His method is based on network analysis, which aims to:

- reveal who knows what in an organization, and
- measure degrees of expertise.

Network analysis has evolved from sociograms that depict the relationships among members of a group. Similarly, network analysis describes transactions between individuals. Stein's network analysis for identifying experts is based on a measurement of information retrieval in an organization. Stein distinguished between two measures: *indegree* and *outdegree*. Outdegree refers to the number of retrieval paths out from an individual, whereas indegree refers to the number of retrieval paths into an individual. Stein's central idea is that experts are the individuals from whom most information is retrieved in an organization. Obviously we also have to take into account the kind of information and—to some extent—the accessibility of the individuals. In the long run, however, as Stein (1992) argued, it is the experts who are most frequently used as sources of knowledge. "Sources that contribute to member success in the choice of appropriate course of action will be retrieved from more often, and these will emerge as source of expertise" (p. 162).

By Stein's method, we can particularly identify relational experts—members within the staff who function as knowledge sources for others in the company. Professionals are identified via their professional affiliation, scientists by their disciplinary affiliation.

Management Expertise

With some reservations, we can consider management as a kind of professional expertise. Reservations concern expertise and decision making. In chapter 3, we argued that in a strict sense experts do not decide; instead they inform or explain. Conflicts with experts can arise from forcing them into a decision. In a company, however, the ultimate decision is made by the management. Managers account for their decisions. Having defined organizations as corporate actors in form of resource pools, we can conceive of management more precisely as directing an organization toward organi-

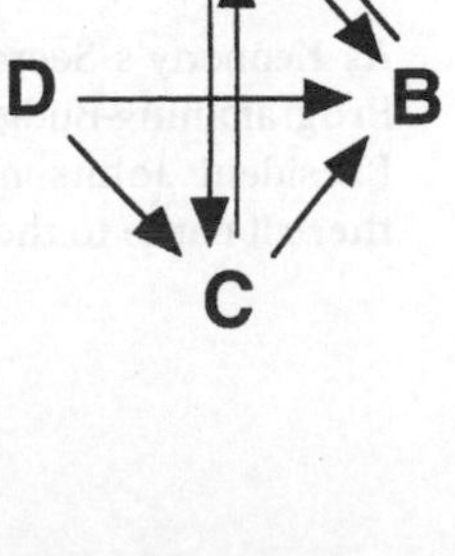

FIG. 8.2. Fictive example of network an trieval by one individual from another. The arrow points to the individual from whom information is retrieved. Obviously in this network, B is the most often frequented source of information—the expert.

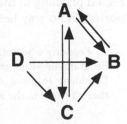

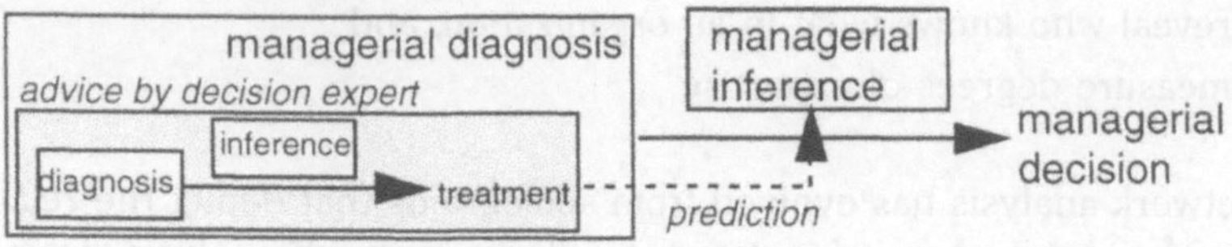

FIG. 8.3. Managerial decision making viewed as a professional task. This figure shows managerial decision making viewed as a professional task (Fig. 8.1) using formal expert advice (gray box) as a decision base for managerial diagnosis. Because formal advice aims to predict what the manager's decision will bring about when based on this formal advice, there is a certain recursivity in managerial decision making.

zational goals (profit, turn-around, image, etc.). Regulating an organization toward organizational goals suffers from the problem of recursivity that we discussed in relation to financial markets: A manager decides on the basis of the expected or factual decisions of subordinates, clients, and the markets that, again, may—more or less directly—depend on the manager's decisions. This is particularly true when experts are involved. Figure 8.3 shows managerial decision making as a professional task that uses formal decision advice—that is, a consultant advises the manager on decision making. Even in this case, it remains the task of the manager to evaluate expert advice. Managers have to particularly take into account on what kind of advice the market competitors act and how this affects his or her own decisions. This inherent recursivity in managerial decision making may be a cause for the lack of predictive theories in management.

8.3 OUTLOOK: DO WE NEED EXPERTS FOR PLANNING?

Robust Planning

In his book *The Rise and Fall of Strategic Planning*, Mintzberg (1994) attacked specific sorts of expert formal decision advice. In detail, he depicted the failure of the planning–programming–budgeting system (PPBS) that directed planning in the McNamara administration and was, as Mintzberg reported, in no way helpful during the Vietnam war.

> As Kennedy's Secretary of Defense, McNamara imposed PPBS [the Planning-Programming-Budgeting-System, HAM] on the military establishment; later President Johnson decreed its use throughout the government and from there it came to the state governments and imitators in other countries. In his

acclaimed book, *The politics of the Budgetary Process*, Aaron Wildavsky summarized the experience succinctly: "PPBS has failed everywhere and at all times" [. . .]. (Mintzberg, 1994, p. 117; see Wildavsky, 1974)

Instead of telling the story of PPBS—that is best told by Mintzberg himself—we try to reconstruct Mintzberg's understanding of planning from the expertise perspective. According to Mintzberg (1994), planning is "a formalized procedure to produce an articulated result, in the form of an integrated system of decisions" (p. 12). Mintzberg's concept of planning is composed of the following elements:

planning = future thinking + controlling + decision making +
integration of decision making + formalization

Formalization is a requirement that defines planning as a professional or science-based task. Without formalization, according to Mintzberg, planning would mean any future-oriented complex decision making. How far we can formalize planning is an open question. In fact, we see that planning essentially is a managerial task. This becomes particularly true when we emphasize the difference between the two tasks:

- to advise someone who has to make a decision, or
- to decide oneself.

When advising decision makers, we should strive for a detailed analysis. This is, after all, experts' work. When deciding, we cannot strive for a detailed analysis due to time and budget constraints. Because management has to deal with humans and markets, managerial decision making is decision making under uncertainty. Thus, managerial decision making is best understood as a question of commitment to decisions and not as a rational choice among different decision alternatives. Only in rare cases do we really know alternatives and their utilities. Therefore, some elements of planning—in Mintzberg's definition—such as *future thinking* and *formalization* can be delegated to experts, whereas other elements such as decision making and controlling should not. Mintzberg (1994) cited a U.S. Army officer who said: "Any damn fool can write a plan. It's the execution that gets you all screwed up" (p. 118; originally cited by Summers, 1981).

In a paper on political planning, Luhmann (1971) defined *planning* as deciding on decisions. That is, on the one hand, another way of saying planning is about determining the way of future action. On the other hand, it shows the specific interaction of prediction and control in planning: When planning, we deliberately decide on what kind of further decisions to make.

Planning literature often refers to this reflexive decision structure by speaking of strategic planning. Most definitions of strategy disappoint, focusing on design or a hierarchy of plans. Let us instead turn to the origins of this word that reveal a specific commitment of the strategists. The Greek word στρατηγόσ (strategos, strategist) referred to the highest military and administrative position—a general. The ancient Greek republic of Athens had 10 elected strategists. As general, a strategist was personally responsible for the success of the troops he commanded. If a battle failed, the strategist had to expect a public trial and severe punishment—even when there seemed neither proof nor probability that the strategist had been guilty of misconduct or misjudgment (see Pritchett, 1974). Thus, being a strategist meant total personal commitment to the final success of all decisions involved, intended or not, by the strategist's management task.

The evaluation of planning is different from the one of expert judgment or expert performance. In particular, the criteria of validity and reliability do not catch the quality of planning. There is no practical use of the validity criterion because planning validity can—if at all—only be measured in the future. Equally, the criterion of reliability is not really relevant to planning. If we repeat the planning process and arrive at a different plan, this does not necessarily put into question our method of planning. Even diverging plans can be equally successful. As to planning, we have to leave the claim of veridicality. Instead, we should resort to criteria for the process such as *robustness*.

Robustness has become a topic in management literature. In their Harvard Business School book on management, Eccles, Nohria, and Berkley (1992) pleaded to refrain from the "obsession with newness" (p. 4), such as "innovative" management or searching for "excellence." They recommended *robust action*—that is, "action that accomplishes short-term objectives while preserving long-term flexibility" (p. 11)—a definition similar to that of *sustainability* (see e.g., Hitchens, Clausen, & Fichter, 1999).

In their book, Eccles, Nohria, and Berkley tried to demonstrate that robust action is essentially in accordance with managerial practice that cannot be based directly on theoretical knowledge. Defending managerial practice, they argued that managers, when consulting management literature, are not as much interested in management concepts, but in examples of managerial experience—"concrete examples of what others have done" (p. 179). From this perspective, managers prefer detailed company stories to assess the relevance of others' experiences as well as their credibility—that is, to assess expertise (by a dispositional attribution of expertise). They wrote:

> The rich and often evocative detail in these popular accounts enables a manager to decide the extent to which the example applies to her own situation. (Eccles et al., 1992, p. 180)

> [T]he credibility of the company or manager that is the source of the story greatly influences how compelling it is seen to be. (loc. cit.)

Similarly, we could consider robust planning as planning that enables us to accomplish short-term objectives while preserving long-term flexibility.

Scenario-Based Planning With the Help of Experts

Let us turn to planning with experts. In chapter 5, we discussed the relationship between planning and prediction. We saw that predicting is confounded with planning in the case of predicting corporate earnings: The prediction is accurate when the company develops according to plan (see Table 5.2). We also saw that expert prediction is more accurate in the short term than in the long term. Thus, managers are well advised to trust expert short-term predictions and to be suspicious of expert long-term forecasts.

If we focus on short-term prediction, managerial planning can profit from integrating expert judgments. This holds, for instance, for scenario-based planning. Scenario construction is a planning tool that was developed as a means for strategic planning by the Rand Corporation in the 1950s and 1960s. In chapter 6, scenario construction was introduced in connection with climate change policies. In general, scenario construction starts with one scenario that is some kind of extrapolation of the present state. This results in the business-as-usual scenario. Then key parameters of the scenario are analyzed. Further scenarios are derived by varying the key parameters. The set of scenarios enables the development and evaluation of strategies. The essence of scenario-based planning is to think in diverging options instead of planning by extrapolation of the present state.

From the point of view of cognitive psychology, scenarios include the use of mental models that contain causal knowledge of this type: "If Parameter X changes, then Parameter Y will change, too." The mental model helps to cognitively simulate future problem development. As discussed in chapter 6, mental simulation uses *joints*—that is, dramatic events or decision points. From a technical point of view, joints are based on the variation of key parameters. We should distinguish between:

- controllable key parameters such as investments, reorganization, and political measurements; and
- uncontrollable key parameters such as business cycles, market growth, and international politics.

Let us turn to an example. Beginning in the 1990s, the European railway companies faced a completely new situation, the railroad system being deregulated. Now there is international competition between former national train operators. The operation of the track system is disconnected from the operation of the trains. Moreover, there is new competition between freight transportation and passenger transportation. New regional private companies arise, exploiting market niches. The railway company managers now have to take into account the following developments:

- technology (e.g., high-speed passenger trains, new coupling system);
- logistics (e.g., freight tracking systems, container systems);
- markets (e.g., competitors, energy costs);
- governmental regulations (e.g., energy taxes, public opinion); and
- European traffic system (e.g., national road pricing, freight traffic flows).

The possible scenarios are defined by combinations of these developments. They describe the kinds of competition the railway companies will have to face and the kinds of client relationships these companies have to build (see e.g., UIC, 1997). For example, the different scenarios focus on:

- significant intermodal competition because railways are only one mode of transportation in transnational traffic;
- the rise of a few transnational train operators that will control the market; and
- a new wave of environmental regulation caused by a change in public opinion after some accidents with severe environmental impacts will have occurred.

Usually professional consultants offer scenario analysis or help in scenario-based planning (e.g., Fahey & Randall, 1998). However, if we distinguish between expert roles, we see that expert advice enters scenario-based managerial planning in two forms. First, experts can help construct and define scenarios; these are decision experts such as consultants or members of the internal organization staff. Second, experts provide causal knowledge for the analysis of the present situation and the scenarios; these are scientific experts or special analysts. Figure 8.4 shows a distribution of work in the case of scenario-based planning, including as experts members of the accounting department. The task of deciding which action to take remains with the managers. Expert advice is necessary for short-term predictions for the present situation as well as for the analysis of the scenarios. Figure 8.4 shows the example of predicting corporate earnings. The short-term predictions of actual earnings can be more or less valid. The short-term predictions of earnings for the scenarios are hypothetical and have a different function: They have to be consistent with the situation depicted by the scenario. In Fig. 8.4, Scenario A depicts business-as-usual and is an

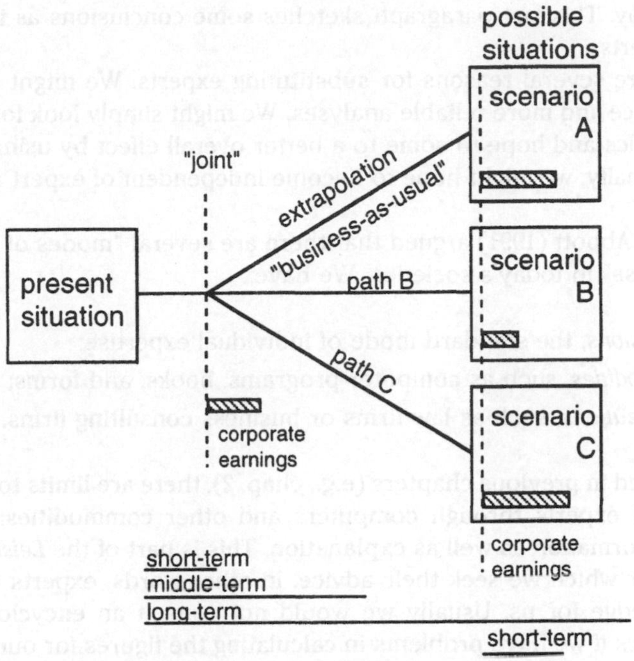

FIG. 8.4. Integrating (short-term) expert judgment in scenario-based managerial planning. The task of deciding which action to take is up to the manager. Short-term expert prediction contributes to the managerial decision making in two ways: first, expert predictions for the actual situation; second, expert predictions within the scenarios.

extrapolation of the present situation. Scenarios B and C depict two alternatives, with Scenario C involving the highest expected earnings. The final valuation of the scenarios is up to the managers. Even if corporate earnings are highest in Scenario C, it does not need to be robust. For instance, a railway company may consider selling its own water power plant because energy costs are presently low due to atomic energy. However, assuming that national governments refrain from atomic energy and energy costs rises, it would be wise to have kept this water power plant. In this way, scenario construction supports robust managerial action.

Substituting Experts

It is not the aim of this book to recommend expert advice for managerial decision making. For instance, scenario construction with experts can be lengthy and costly. In case we are about to use experts and expertise, *The Social Psychology of Expertise* can show how to make use of them in the most efficient way. This last paragraph sketches some conclusions as to substituting experts.

There are several reasons for substituting experts. We might hope for better advice and more reliable analyses. We might simply look for a variation of advice and hope to come to a better overall effect by using several experts. Finally, we might hope to become independent of expert advice at all.

Andrew Abbott (1991) argued that there are several "modes of embodying expertise" in today's societies. We have:

- *professions*, the standard mode of individual expertise;
- *commodities*, such as computer programs, books, and forms; and
- *organizations*, such as law firms or business consulting firms.

As discussed in previous chapters (e.g., chap. 2), there are limits to the substitution of experts through computers and other commodities. Experts provide information as well as explanation. This is part of the *Leistung* (see chap. 3) for which we seek their advice. In other words, experts interpret the knowledge for us. Usually we would not consult an encyclopedia of mathematics if we have problems in calculating the figures for our tax declaration. The future will show how tutoring computer systems can overcome these problems (see Mieg, 1993).

Instead of individual experts, we can also consult organizations. This includes not only business consulting firms and universities, but also associations of all kinds (e.g., consumer councils). Organizations store information and expertise and are able to solve complex problems. The *Leistung* of an expert (effect and efficiency), for the sake of which experts are used, is the relatively fast utilization of the expert's compressed experience any reasonable person could make if he or she had enough time to do so. Organizations can serve the same purpose as individual experts. Their size is their advantage: In contrast to a single expert, a single organization offers the same service at the same time to several clients. Moreover, organizations as corporate actors have one important feature in common with individual experts: They are potential objects of trust. Trust is a necessary requirement of expert services. For distrusting an expert would also mean distrusting the information and explanations provided by that expert. In this way, we trust individuals and we trust (or distrust) organizations. Due to its size, an organization has a marketing advantage: It can more easily become a brand name for qualitative expertise than a single expert can.

However, when we consult a company as a source of knowledge or know-how, the internal processes in this particular company involve internal experts and knowledge transactions. We can assume that there are the same organizational role conflicts with experts as we discussed them in general in chapter 4. These role conflicts focusing on questions like: Who is competent? Who is motivated? Who is accountable? are invisible to us.

Finally, there is one important advantage for individual experts: They can be more easily substituted than organizations. We can engage and dismiss experts, and we can acquire and sell organizations. Yet the transaction costs are quite different. In particular, by personnel management, a company can arrange its expertise portfolio and knowledge base. The expert role approach that is comprised in Table 8.2. provides concepts for improving the management of distributed work, in particular when assigning special tasks to individuals with specific expertise—the experts.

Bibliographical Notes

These notes are in addition to the references made in the text.

CHAPTER 1

The psychology of expertise. Besides the book by Chi, Glaser, and Farr (1988), see the works by Ericsson (e.g., Ericsson, 1996; Ericsson & Charness, 1994, 1997; Ericsson & Smith, 1991b); Feltovich, Ford, and Hoffman (1997); Hoffman (1992); Hoffman, Shadbolt, Burton, and Klein (1995); and Wright and Bolger (1992). An interdisciplinary view on expertise is provided by Williams, Faulkner, and Fleck (1998). This book touches on several points relevant to a social psychology of expertise, such as the fact that "what counts as expertise is socially contingent" (p. 8) or the "tradeability" of expertise (chap. 7). As to research in Germany, see, for example, Bromme (1992) and Gruber and Ziegler (1996).

CHAPTER 2

Bootstrapping. The result that statistical diagnoses outperformed clinicians' judgments (Goldberg, 1969, 1970) is more generally known as *bootstrapping* (see e.g., Bolger & Wright, 1992; Blattberg & Hoch, 1990; Dawes & Corrigan, 1974): Using a (regression) model of our own diagnostic judgments, anyone of us can "pull him/herself up by his/her own bootstraps" (Dawes & Corrigan, 1974) and perform better than without a model.

Experts, expert systems, professional work, and uncertainty. A comprehensive overview of experts and uncertainty–from a decision-science perspective–is provided by Cooke (1991). Dreyfus and Dreyfus (1986) argued from a philosophical point of view against the feasibility of simulating human cognition. As Mieg (1993) argued, they chose the wrong example–chess. Chess is a well-defined problem (cf. e.g., McCarthy, 1956); no wonder that an engineering solution for computing chess, such as Deep Blue, was able to dominate our human chess masters (Chabris, 1997). Winograd and Flores (1986) argued that working as an expert in an organization requires situational pre-understanding that cannot be formalized (relevance decisions according to Mieg, 1993). Gerrity, Earp, DeVellis, and Light (1992) discussed uncertainty in the everyday work of physicians. As to expert systems, see, for example, Hayes-Roth, Waterman, and Lenat (1983) and Weizenbaum (1976).

Rationality. For a more extended discussion see, for instance, Allais (1953); Daston (1988); Gigerenzer (1996); Gigerenzer and Goldstein (1996); Grandy and Warner (1988); Kahneman, Slovic, and Tversky (1982); Mieg (1993); and Rescher (1988).

The sociology of the professions. For a more comprehensive discussion of the sociology of professions (besides Abbott, 1988, 1991) see, for example, Carr-Saunders and Wilson (1933), Evetts (1999), Freidson (1970a, 1970b, 1983, 1986a, 1986b, 1999), Johnson (1967, 1977), Larson (1977), Parsons (1951, 1968), and Rueschemeyer (1983, 1986). See also *The Reflective Practitioner* by Schön (1991).

CHAPTER 3

Mead. We base our description mainly on Mead's *Mind, Self, and Society,* which was posthumously edited by Charles W. Morris, and on one of Mead's last and most complex papers on that topic: *The Genesis of the Self and Social Control* that appeared in 1924/1925. We cite *Mind, Self and Society* in the edition of selected papers by Anselm Strauss (Mead, 1977) that unfortunately does not include the paper on self- and social control.

Attribution theory. There are several lines of research in attribution theory. Clarification of the attribution matrix as in Table 3.1 is largely due to Weiner (1986). Weiner (1986, 1995) investigated the attribution of responsibility, especially in combination with affect and emotion. Kelley (1967, 1973) provided a model for the attribution process. Medcof (1990) presented an "integrative model of attribution processes," called PEAT (probability, expectations, attribution theory), that reinterprets attribution theory in terms of probability beliefs. Tetlock and Levi (1982) discussed "the inconclusiveness of the cognition-motivation debate" in attributional effects. As to the general matter of accountability, see Semin and Manstead (1983).

Trust. As to the concept of trust, see also Luhmann (1973, 1988); for an overview and discussion, see Lewis and Weigert (1985) or Mayer, Davis, and Schoorman (1995).

The philosophy of science. Besides Fleck (1935/1979) and Kuhn (1962), see Callebaut and Pinxten (1987), Carnap (1947), Stegmüller (1979), and von Wright (1971).

CHAPTER 4

Legitimizing uses of experts. As to the different uses of experts, see Freidson (1986b), Headrick (1992), Janis (1972), Jasanoff (1994, 1995), Martin (1996), Nowotny (1979), and Peters (1995).

Metacognitive knowledge. The parallel between why-explanations and metacognitive knowledge is not as obvious and convincing as the parallels between what-explanations and declarative knowledge or between how-explanations and procedural knowledge, respectively. Metcalfe (1993) investigated metacognition as a kind of metamemory that assesses the novelty of information. "Even though they [people] are unable to recall a particular fact or event, they can assess quite accurately how likely it is that they will be able to recall or recognize it some time later" (p. 3). See also Flavell and Wellman (1977).

Experts and groups. There is, of course, research on this topic in social psychology besides Wegner and Stasser. For an overview, see Levine, Resnick, and Higgins (1993). See also Janis (1972), Shulman and Carey (1984), Mullen and Goethals (1987), and Galegher, Kraut, and Egido (1990).

CHAPTER 5

Variance, semivariance, covariance in portfolio theory. As Markowitz (1991) admitted, semivariance would be more "plausible than variance as a measure of risk" (p. 476). However, both variance and semivariance concepts of risk have been shown not to agree with von Neumann and Morgenstern's (1944) axioms of rational preferences except under rather limited conditions (Levy & Markowitz, 1979; Levy & Sarnat, 1984). We could also argue that the true concept of risk within portfolio selection theory is not variance but covariance. From a statistical point of view, we cannot leave out modeling the covariances: "The covariance structure that drives the Markowitz model is abstract, complex, and non-observable. There is absolutely nothing in the background or training of a security analyst (or anyone else) that would qualify or require him or her to quantify, or otherwise render, a subjective judgment about a covariance statistic" (Frankfurter & Phillips, 1995, p. 66).

Theory and research on processes in financial markets. As to some market deviations (from the CAPM point of view), see, for example, Fama and French (1992), who showed that the alleged positive correlation between risk and return is confounded with a size effect. Within the groups of small stocks and large stocks, respectively, there is a negative correlation between risk and return. Fama is the most prominent researcher on efficient markets (Fama, 1970, 1976, 1991). As to (market) overreaction, see DeBondt and Thaler (1985, 1987, 1990), who also discuss other anomalies. For experimental research on market behavior, see also Sarin and Weber (1993) or Weber and Camerer (1998). As to market volatility, see Shiller (1991, 2000). As to market bubbles, see Garber (1989, 1990).

Financial expertise, the psychology of finance. Thaler (1993) and Shefrin (1999) summarized research on behavioral finance. The results by Stael von Holstein (1972) were partly replicated by Yates, McDaniel, and Brown (1991). They focused on graduate and undergraduate students in finance. They found an inverted expertise effect: "Undergraduate students were more accurate than graduate subjects" (p. 60). For interviews with traders, see Schwager (1989, 1992). As to the case of Nick Leeson and Barings Bank, see Leeson and Whitley (1996). As to the case of the "junk-bond king" Michael Milken and the crash of Drexel Burnham Lambert, see Bally (1991) or Zey (1993). Drexel Burnham Lambert was a Wall Street firm that pioneered in the junk-bond market. Junk-bonds are high-yield bonds of companies that are below standard investment grades as measured by the credit-rating agencies Moody's and Standard & Poor's.

CHAPTER 6

Climate change research. Literature on climate change abounds. The case study focuses on the IPCC reports from 1990 to 1996 (Houghton, Callander, & Varney, 1992; Houghton, Jenkins, & Ephraums, 1990; Houghton et al., 1995, 1996). Martens and Rotmans (1999) provided an integrative perspective on climate change issues, including the integrated assessment approach (Rotmans & van Asselt, 1999). Boehmer-Christiansen (1994a, 1994b) depicted the history of the global climate change policy. In this context, see also Jepma and Munasinghe (1998). The work of the World Climate Research Programme, founded in 1979 by the World Meteorologial Organization (WMO), is described by Kondratyev and Cracknell (1998). As to an assessment of the impact of climate change on the U.S. economy, see Mendelsohn and Neumann (1999). They totally contradicted the IPCC economic assessment, being more optimistic. As to a regional assessment of climate for the Alpine region, see Cebon, Dahinden, Davies, Imboden, and Jaeger (1998). For information on the UN Framework Convention on Climate Change (UNFCCC) and the Kyoto Protocol, see Grubb (1999) and www.unfccc.org.

System experts in environmental planning—The ETH-UNS-Case Studies. The term *system expert* has been created in the context of the ETH-UNS case studies. The integrated assessment approach does not speak of system or lay experts. On the contrary, the papers on integrated assessment attribute expert to science and scientists (cf. Kasemir & Jaeger, 1997, p. 1; Kasemir, van Asselt, et al., 1997, p. 71). The ETH-UNS case studies are teaching projects at the Swiss Federal Institute of Technology (ETH). They are concerned with real-case environmental planning (e.g., regional development). For a description of the ETH-UNS Case Studies, see Mieg (1996); Scholz, Mieg, and Weber (1997); and Scholz and Tietje (in press). Mieg (2000) discussed the use of experts in ETH-UNS case studies.

CHAPTER 7

Formal decision-making advice. As to the theory of decision making and its application, see Cooke (1991); Kleindorfer, Kunreuther, and Schoemaker (1993); von Winterfeldt and Edwards (1986); and Yates (1990).

CHAPTER 8

The expert role approach in environmental planning. For experiences with the implementation of the expert role approach in environmental planning, see Mieg (2000). Project management in environmental planning should be based on the principle of modular integration (Mieg, Scholz, & Stünzi, 1996), which is partly integrating external expertise (experts, organizations) while preserving differences in objectives (of different stake holders or involved firms).

References

Abbott, A. (1988). *The system of professions*. Chicago: The University of Chicago Press.

Abbott, A. (1991). The future of professions. *Research in the Sociology of Organizations*, *8*, 17–42.

Abraham, H. J. (1993). *The judicial process* (6th ed.). New York: Oxford University Press.

Agnew, N. M., Ford, K. M., & Hayes, P. J. (1997). Expertise in context: Personally constructed, socially selected and reality-relevant? In P. J. Feltovich, K. M. Ford, & R. R. Hoffman (Eds.), *Expertise in context: Human and machine* (pp. 219–244). Menlo Park, CA: AAAI.

Allais, M. (1953). Le comportement de l'homme rationnel devant le risque: Critique des postulats et axiomes de l'école americaine [Human rational behavior in connection with risk: Criticism of the postulates and axioms of the American school]. *Econometrica*, *21*(4), 503–546.

Allen, R. J., & Miller, J. S. (1993). The common law theory of experts: Deference or education? *Northwestern University Law Review*, *87*(4), 1131–1147.

Anderson, J. R. (1983). Methodologies for studying human knowledge. *Behavioral and Brain Sciences*, *10*, 467–505.

Anderson, J. R. (1985). *Cognitive psychology and its implications* (2nd ed.). San Francisco: Freeman.

Anderson, J. R. (1990). *The adaptive character of thought*. Hillsdale, NJ: Lawrence Erlbaum Associates.

Anderson, J. R. (1995). *Learning and memory*. New York: Wiley.

Andreassen, P. B. (1987). On the social psychology of the stock market: Aggregate attributional effects and the regressiveness of prediction. *Journal of Personality and Social Psychology*, *53*(3), 490–496.

Andreassen, P. B. (1988). Explaining the price-volume relationship: The difference between price changes and changing prices. *Organizational Behavior and Human Decision Processes*, *41*, 371–389.

Andreassen, P. B. (1990). Judgmental extrapolation and market overreaction: On the use and disuse of news. *Journal of Behavioral Decision Making*, *3*, 153–174.

Aranya, N., & Ferris, K. R. (1984). A reexamination of accountants' organizational-professional conflict. *The Accounting Review*, *LIX*(1), 1–15.

Aranya, N., Pollock, J., & Amernic, J. (1981). An examination of professional commitment in accounting. *Accounting, Organizations and Society*, *6*(4), 271–280.

Auerbach, R. D. (1985). *Money, banking, and financial markets*. New York: Macmillan.

Baer, W. C. (1986). Expertise and professional standards. *Work and Occupations, 13*, 532–552.

Bailey, F. (1991). *The junk bond revolution*. London: Mandarin.

Barber, B. M., & Loeffler, D. (1993). The "dartboard" column: Second-hand information and price pressure. *Journal of Financial and Quantitative Analysis, 29*(2), 273–284.

Baumbach, A., Lauterbach, W., Albers, J., & Hartmann, P. (1993). *Zivilprozeßordnung* (51. Aufl.) [German Code on Civil Procedures]. München: C. H. Beck.

Becker, G. S. (1993). *Human capital* (3rd ed.). Chicago: The University of Chicago Press.

Bierbrauer, G. (1979). Why did he do it? Attribution of obedience and the phenomenon of dispositional bias. *European Journal of Social Psychology, 9*, 67–84.

Black, F., & Scholes, M. (1973). The pricing of options and corporate liabilities. *Journal of Political Economy, 81*(3), 637–654.

Blattberg, R. C., & Hoch, S. J. (1990). Database models and managerial intuition. *Management Science, 36*(8), 887–899.

Boehmer-Christiansen, S. (1994a). Global climate protection policy: The limits of scientific advice. Part 1. *Global Environmental Change, 4*(2), 140–159.

Boehmer-Christiansen, S. (1994b). Global climate protection policy: The limits of scientific advice. Part 2. *Global Environmental Change, 4*(3), 185–200.

Bolger, F., & Wright, G. (1992). Reliability and validity in expert judgment. In G. Wright & F. Bolger (Eds.), *Expertise and decision support* (pp. 47–76). New York: Plenum.

Bromme, R. (1992). *Der Lehrer als Experte* [The teacher as expert]. Bern: Huber.

Brown, R. V. (1989). Toward a prescriptive science and technology of decision aiding. *Annals of Operations Research, 19*, 467–483.

Brown, R. V., Larichev, O., Flanders, N., & Andreyeva, E. (1995, August). *Decision aids for land-use permitting: An Alaska oil case*. Paper presented at SPUDM 15 (Subjective Probability Utility and Decision Making), Jerusalem.

Buchanan, B. G., & Shortliffe, E. H. (Eds.). (1984). *Rule-based expert systems*. Reading: Addison-Wesley.

Buckingham-Hatfield, S., & Percy, S. (1999). *Constructing local environmental agendas*. London: Routledge.

Burk, J. (1988). *Values in the marketplace*. Berlin: deGruyter.

Callebaut, W., & Pinxten, R. (Eds.). (1987). *Evolutionary epistemology*. Dordrecht, Holland: D. Reidel.

Campbell, D. T. (1960). Blind variation and selective retention in creative thought as in other knowledge processes. *Psychological Review, 67*(6), 380–400.

Capstaff, J., Paudyal, K., & Rees, W. (1995). The accuracy and rationality of earnings forecasts by UK analysts. *Journal of Business Finance & Accounting, 22*(1), 67–85.

Carnap, R. (1928). *Der logische Aufbau der Welt* [The logical structure of the world]. Hamburg: Meiner.

Carnap, R. (1947). *Meaning and necessity*. Chicago: The University of Chicago Press.

Carr-Saunders, A. M., & Wilson, P. A. (1933/1964). *The professions*. London: Frank Cass.

Cebon, P., Dahinden, U., Davies, H., Imboden, D. M., & Jaeger, C. C. (Eds.). (1998). *Views from the alps: Regional perspectives on climate change*. Cambridge, MA: The MIT Press.

Cecil, J. S., & Willging, T. E. (1993). *Court-appointed experts: Defining the role of experts appointed under Federal Rule of Evidence 706*. Federal Judicial Center.

Chabris, C. (1997). Brave new chess world. *American Chess Journal*. Published electronically: www.h3.org/pub/acj/extra/Chabris/Chabris04.html.

Charness, N. (1991). Expertise in chess: The balance between knowledge and search. In K. A. Ericsson & J. Smith (Eds.), *Toward a general theory of expertise* (pp. 39–63). Cambridge, MA: Cambridge University Press.

Chase, W. G., & Simon, H. A. (1973). The mind's eye in chess. In W. G. Chase (Ed.), *Visual information processing* (pp. 215–281). New York: Academic Press.

Chi, M. T. H., Glaser, R., & Farr, M. J. (Eds.). (1988). *The nature of expertise*. Hillsdale, NJ: Lawrence Erlbaum Associates.

Chi, M. T. H., Glaser, R., & Rees, E. (1982). Expertise in problem solving. In R. Sternberg (Ed.), *Advances in the psychology of human intelligence* (pp. 17–76). Hillsdale, NJ: Lawrence Erlbaum Associates.

Clancey, W. J. (1997a). *Situated cognition: On human knowledge and computer representations*. Cambridge, UK: Cambridge University Press.

Clancey, W. J. (1997b). The conceptual nature of knowledge, situations, and activity. In P. M. Feltovich, K. M. Ford, & R. R. Hoffman (Eds.), *Expertise in context: Human and machine* (pp. 247–291). Menlo Park, CA: AAAI.

Clayman, M. (1987, May–June). "In search of excellence": The investor's viewpoint. *Financial Analysts' Journal*, pp. 54–63.

Climate Research Board. (1979). *Carbon dioxide and climate: Scientific assessment*. Washington, DC: National Academy of Sciences.

Coleman, J. S. (1974). *Power and the structure of society*. New York: W. W. Norton.

Collins, H. M. (1985). *Changing order*. London: Sage.

Cooke, R. M. (1991). *Experts in uncertainty*. New York: Oxford University Press.

Corwin, R. G. (1961). The professional employee: A study of conflict in nursing roles. *The American Journal of Sociology, LXVI*, 604–615.

Crozier, M. (1963). *Le phénomène bureaucratique*. Paris: Seuil.

Crozier, M. (1964). *The bureaucratic phenomenon*. Chicago: The University of Chicago Press.

Daston, L. (1988). *Classical probability in the Enlightenment*. Princeton: Princeton University Press.

Davenport, T. H. (1993). *Process innovation*. Boston, MA: Harvard Business School Press.

Dawes, R. M., & Corrigan, B. (1974). Linear models in decision-making. *Psychological Bulletin, 81*, 95–106.

De Bondt, W. F. M., & Thaler, R. H. (1985). Does the stock market overreact? *The Journal of Finance, XL*(3), 793–808.

De Bondt, W. F. M., & Thaler, R. H. (1987). Further evidence on investor overreaction and stock market seasonality. *The Journal of Finance, XLII*(3), 557–581.

De Bondt, W. F. M., & Thaler, R. H. (1990). Do security analysts overreact? *AEA Papers and Proceedings, 80*(2), 52–57.

deGroot, A. D. (1965). *Thought and choice in chess*. The Hague: Mouton.

den Elzen, M. (1994). *Global environmental change*. Utrecht: International Books.

Dickinson, R. E. (1986). How will climate change? In B. Bolin, B. R. Döös, J. Jäger, & R. A. Warrick (Eds.), *The greenhouse effect, climatic change, and ecosystems* (pp. 207–270). Chichester: Wiley.

Dietz, T. M., & Rycroft, R. W. (1987). *The risk professionals*. New York: Russell Sage Foundation.

Dingwall, R., & Lewis, P. (Eds.). (1983). *The sociology of the professions*. London: Macmillan.

Dippel, K. (1986). *Die Stellung des Sachverständigen im Strafprozeß* [The legal position of experts in criminal cases]. Heidelberg: R.v.Decker's.

Dreyfus, H. L., & Dreyfus, S. E. (1986). *Mind over Machine*. Oxford: Blackwell.

Dürrenberger, G. (1997). *Focus groups in Integrated Assessment: A manual for a participatory tool* (ULYSSES Report No. WP-97-2). Darmstadt University of Technology, Center for Interdisciplinary Studies in Technology.

Eccles, R. G., Nohria, N., & Berkley, J. D. (1992). *Beyond the hype: Rediscovering the essence of management*. Boston, MA: Harvard Business School Press.

Einhorn, H. J., & Hogarth, R. M. (1978). Confidence in judgment: Persistence of the illusion of validity. *Psychological Review, 85*(5), 395–416.

Einzig, P. (1966). *Primitive money* (2nd ed.). Oxford: Pergamon.

Elstein, A. S., Shulman, L. S., & Sprafka, S. A. (1978). *Medical problem solving*. Cambridge, MA: Harvard University Press.

Ericsson, A. K. (Ed.). (1996). *The road to excellence: The acquisition of expert performance in the arts and sciences, sports, and games*. Mahwah, NJ: Lawrence Erlbaum Associates.

Ericsson, A. K., & Charness, N. (1994). Expert performance: Its structure and acquisition. *American Psychologist, 49*(8), 725–747.

Ericsson, A. K., & Charness, N. (1997). Cognitive and developmental factors in expert performance. In P. J. Feltovich, K. M. Ford, & R. R. Hoffman (Eds.), *Expertise in context: Human and machine* (pp. 3–41). Menlo Park, CA: AAAI.

Ericsson, K. A., Krampe, R. T., & Tesch-Römer, C. (1993). The role of deliberate practice in the acquisition of expert performance. *Psychological Review, 100*(3), 363–406.

Ericsson, K. A., & Smith, J. (1991a). Prospects and limits of the empirical study of expertise: An introduction. In K. A. Ericsson & J. Smith (Eds.), *Toward a general theory of expertise* (pp. 1–38). Cambridge, MA: Cambridge University Press.

Ericsson, K. A., & Smith, J. (Eds.). (1991b). *Toward a general theory of expertise*. Cambridge, MA: Cambridge University Press.

European Commission. (Ed.). (1996). *The markets for electronic information services in the European economic area* (Report No. DG XIII/E). Luxembourg: Author.

Evetts, J. (1999). Professions: Changes and continuities. *International Review of Sociology, 9*, 75–85.

Fahey, L., & Randall, R. M. (Eds.). (1998). *Learning from the future: Competitive foresight scenarios*. Chichester: Wiley.

Fama, E. F. (1970). Efficient capital markets: A review of theory and empirical work. *The Journal of Finance, 25*(5), 383–417.

Fama, E. F. (1976). *Foundations of finance*. New York: Basic Books.

Fama, E. F. (1991). Efficient capital markets II. *The Journal of Finance, XLVI*(5), 1575–1617.

Fama, E. F., & French, K. R. (1992). The cross-section of expected stock returns. *The Journal of Finance, XLVII*(2), 383–417.

Feigenbaum, E. A., & McCorduck, P. (1984). *The fifth generation*. London: Michael Josep H.

Feltovich, P. J., Ford, K. M., & Hoffman, R. R. (Eds.). (1997). *Expertise in context: Human and machine*. Menlo Park, CA: AAAI.

Flavell, J. H., & Wellman, H. M. (1977). Metamemory. In R. V. Kail & R. W. Hagen (Eds.), *Perspectives on the development of memory and cognition* (pp. 3–33). Hillsdale, NJ: Lawrence Erlbaum Associates.

Fleck, L. (1935/1979). *Genesis and development of a scientific fact*. Chicago: The University of Chicago Press. (Original work in German, published 1935)

Folkerts-Landau, D., & Ito, T. (1995). *International capital markets*. Washington: International Monetary Fund.

Frankfurter, G. M., & Phillips, H. E. (1995). *Forty years of normative portfolio theory*. Greenwich, CT: JAI.

Freidson, E. (1970a). *Profession of medicine: A study of the sociology of applied knowledge*. New York: Dodd, Mead and Co.

Freidson, E. (1970b). *Professional dominance: The structure of medical care*. Chicago: Aldine-Atherton.

Freidson, E. (1983). The theory of professions: State of the art. In R. Dingwall & P. Lewis (Eds.), *The sociology of the professions* (pp. 19–37). London: Macmillan.

Freidson, E. (1986a). The medical profession in transition. In L. H. Aiken & D. Mechanic (Eds.), *Applications of social science to clinical medicine and health politics*. New Brunswick, NJ: Rutgers University Press.

Freidson, E. (1986b). *Professional powers*. Chicago: The University of Chicago Press.

Freidson, E. (1999). Theory of professionalism: Method and substance. *International Review of Sociology, 9*, 117–129.

Freyhof, H., Gruber, H., & Ziegler, A. (1992). Expertise and hierarchical knowledge in representation in chess. *Psychological Research, 54*, 32–37.

Galegher, J., Kraut, R. E., & Egido, C. (Eds.). (1990). *Intellectual teamwork*. Hillsdale, NJ: Lawrence Erlbaum Associates.

Garber, P. M. (1989). Tulipmania. *Journal of Political Economy, 97*(3), 535–560.

Garber, P. M. (1990). Famous first bubbles. *Journal of Economic Perspectives, 4*(2), 35–54.

Gerrity, M. S., Earp, J. A. L., DeVellis, R. F., & Light, D. W. (1992). Uncertainty and professional work: Perceptions of physicians in clinical work. *American Journal of Sociology, 97*(4), 1022–1051.

Gibbons, M., Limoges, C., Nowotny, H., Schwartzman, S., Scott, P., & Trow, M. (1994). *New production of knowledge: Dynamics of science and research in contemporary societies*. London: Sage.

Gigerenzer, G. (1996). Rationality: Why social context matters. In P. B. Baltes & U. M. Staudinger (Eds.), *Interactive minds* (pp. 319–346). Cambridge, UK: University Press.

Gigerenzer, G., & Goldstein, D. G. (1996). Reasoning the fast and frugal way: Models of bounded rationality. *Psychological Review, 103*(4).

Gigerenzer, G., & Murray, D. J. (1987). *Cognition as intuitive statistics*. Hillsdale, NJ: Lawrence Erlbaum Associates.

Goffman, E. (1959). *The presentation of self in everyday life*. New York: Anchor Books.

Goldberg, L. R. (1969). The search for configural relationships in personality assessment: The diagnosis of psychosis vs. neurosis from the MMPI. *Multivariate Behavioral Research, 4*, 523–536.

Goldberg, L. R. (1970). Man vs. model of man: A rationale, plus some evidence, for a method of improving clinical inference. *Psychological Bulletin, 73*, 422–432.

Goode, W. J. (1957). Community within a community: The professions. *American Sociological Review, 22*, 194–200.

Goode, W. J. (1969). The theoretical limits of professionalization. In A. Etzioni (Ed.), *The semi-professions and their organizations* (pp. 266–313). New York: The Free Press.

Goodman, N. (1954). *Fact, fiction and forecast*. London: University of London/Athlone Press.

Graham, B., & Dodd, D. (1934/1951). *Security analysis* (3rd ed.). New York: McGraw-Hill.

Grandy, R. E., & Warner, R. (Eds.). (1988). *Philosophical grounds of rationality*. Oxford: Clarendon.

Groner, R., Groner, M., & Bischof, W. F. (1983). *Methods of heuristics*. Hillsdale, NJ: Lawrence Erlbaum Associates.

Grover, S. L. (1993). Why professionals lie: The impact of professional role conflict on reporting accuracy. *Organizational Behavior and Human Decision Processes, 55*, 251–272.

Grubb, M. (1999). *The Kyoto Protocol*. London: The Royal Institute of International Affairs.

Gruber, H., & Ziegler, A. (Hrsg.). (1996). *Expertiseforschung* [Research on expertise]. Opladen: Westdeutscher Verlag.

Gunz, H. P., & Gunz, S. P. (1994). Professional/organizational commitment and job satisfaction for lawyers. *Human Relations, 47*(7), 801–828.

Hacker, W. (1992). *Expertenkönnen* [Expert know-how]. Göttingen: Verlag für Angewandte Psychologie.

Hafif, H. (1973). Jury selection in light of expected expert testimony. In G. W. Holmes (Ed.), *Experts in litigation* (pp. 25–33). Ann Arbor: Institute of Continuing Legal Education.

Hall, R. H. (1968). Professionalization and bureaucratization. *American Sociological Review, 33*, 92–104.

Hammer, M., & Champy, J. (1993). *Reengineering the corporation*. New York: HarperCollins.

Harnad, S. (Ed.). (1987). *Categorical perception*. Cambridge, MA: Cambridge University Press.

Haugen, R. A. (1995). *The new finance*. Englewood Cliffs, NJ: Prentice-Hall.

Hayes, J. R. M. (1952, January–June). Memory span for several vocabularies as a function of vocabulary size. *In Quarterly Progress Report*. Cambridge, MA: Massachusetts Institute of Technology, Acoustics Laboratory. (Reference from Miller, 1956)

Hayes-Roth, F., Waterman, D. A., & Lenat, D. B. (Eds.). (1983). *Building expert systems*. Reading: Addison-Wesley.

Headrick, T. E. (1992). Expert policy analysis and bureaucratic politics: Searching for the causes of the 1987 stock market crash. *Law & Policy, 14*(4), 313–335.
Heider, F. (1958). *The psychology of interpersonal relations*. Hillsdale, NJ: Lawrence Erlbaum Associates.
Hitchens, D. M. W. N., Clausen, J., & Fichter, K. (Eds.). (1999). *International environmental management benchmarks*. Berlin: Springer.
Hitzler, R. (1994). Wissen und Wesen des Experten [Knowledge and nature of experts]. In R. Hitzler, A. Honer, & C. Maeder (Hrsg.), *Expertenwissen* (pp. 13–30). Opladen: Westdeutscher Verlag.
Hoffman, R. R. (Ed.). (1992). *The psychology of expertise*. New York: Springer.
Hoffman, R. R., Feltovich, P. J., & Ford, K. M. (1997). A general framework for conceiving of expertise and expert systems in context. In P. J. Feltovich, K. M. Ford, & R. R. Hoffman (Eds.), *Expertise in context: Human and machine* (pp. 543–580). Menlo Park, CA: AAAI.
Hoffman, R. R., Shadbolt, N. R., Burton, A. M., & Klein, G. (1995). Eliciting knowledge from experts: A methodological analysis. *Organizational Behavior and Human Decision Processes, 62*(2), 129–158.
Hogarth, R. M., & Makridakis, S. (1987). Forecasting and planning: An evaluation. In S. Makridakis & S. C. Wheelwright (Eds.), *The handbook of forecasting* (pp. 539–570). New York: Wiley.
Houghton, J. T., Callander, B. A., & Varney, S. K. (1992). *Climate change 1992: The supplementary report to the IPCC scientific assessment*. Cambridge, UK: University Press.
Houghton, J. T., Jenkins, G. J., & Ephraums, J. J. (1990). *Climate change: The IPCC scientific assessment*. Cambridge, UK: University Press.
Houghton, J. T., Meira Filho, L. G., Bruce, J., Lee, H., Callander, B. A., Haites, E., Harris, N., & Maskell, K. (1995). *Climate change 1994: Radiative forcing of climate change and evaluation of the IPCC IS92 emission scenarios*. Cambridge, UK: University Press.
Houghton, J. T., Meira Filho, L. G., Callander, B. A., Harris, N., Kattenberg, A., & Maskell, K. (1996). *Climate change 1995: The science of climate change*. Cambridge, UK: University Press.
Huber, P. (1993). Galileo's revenge: Junk science in the courtroom. New York: Basic.
Hughes, E. C. (1965). Professions. In K. S. Lynn (Ed.), *The professions in America* (pp. 1–14). Boston: Houghton Mifflin.
Hughes, H. D. (1917). An interesting corn seed experiment. *The Iowa Agriculturist, 17*, 424–425.
Hutchins, E. (1995). *Cognition in the wild*. Cambridge, MA: The MIT Press.
IPCC (Ed.). (1990). = Houghton et al. 1990.
IPCC (Ed.). (1992). = Houghton et al. 1992.
IPCC (Ed.). (1995). = Houghton et al. 1995.
IPCC (Ed.). (1996). = Houghton et al. 1996.
Jacoby, S., & Gonzales, P. (1991). The constitution of expert-novice in scientific discourse. *Issues in Applied Linguistics, 2*(2), 149–181.
Janis, I. L. (1972). *Victims of groupthink*. Boston: Houghton Mifflin.
Jaques, E. (1976). *A general theory of bureaucracy*. London: Heinemann.
Jasanoff, S. (1994). *The fifth branch*. Cambridge, MA: Harvard University Press.
Jasanoff, S. (1995). *Science at the bar: Science and technology in American law*. Cambridge, MA: Harvard University Press.
Jepma, C. J., & Munasinghe, M. (1998). *Climate change policy*. Cambridge, UK: University Press.
Jessnitzer, K., & Frieling, G. (1992). *Der gerichtliche Sachverständige* (10. Aufl.) [The expert in court]. Köln: Carl Heymanns.
Johnson, T. (1967). *Professions and power*. London: Macmillan.
Johnson, T. (1977). The professions in the class structure. In R. Scase (Ed.), *Industrial society* (pp. 93–108). London: George Allen & Unwin.
Johnson-Laird, P. N. (1983). *Mental models*. Cambridge: Cambridge University Press.
Jones, E. E., & Nisbett, R. (1971). *The actor and the observer: Divergent perceptions of the causes of behavior*. Morristown, NJ: General Learning Press.

Jones-Lee, M., & Loomes, G. (2000). Private values and public policy. In E. U. Weber, J. Baron, & G. Loomes (Eds.), *Conflict and tradeoffs in decision making.* New York: Cambridge University Press.

Jungermann, H., & Thüring, M. (1987). The use of causal knowledge for inferential reasoning. In J. L. Mumpower, L. D. Phillips, O. Renn, & V. R. R. Uppuluri (Eds.), *Expert judgment and expert systems* (pp. 131–146). Berlin: Springer.

Jungermann, H., & Thüring, M. (1988). The labyrinth of experts' minds: Some reasoning strategies and their pitfalls. *Annals of Operations Research, 16,* 117–130.

Kahn, H. (1960). *On thermonuclear war.* New York: The Free Press.

Kahn, H., & Wiener, A. J. (1967). *The year 2000.* New York: Macmillan.

Kahneman, D. (1991). Judgment and decision making: A personal view. *Psychological Science, 2,* 142–145.

Kahneman, D., Slovic, P., & Tversky, A. (Eds.). (1982). *Judgment under uncertainty: Heuristics and biases.* Cambridge, MA: Cambridge University Press.

Kahneman, D., & Tversky, A. (1973). On the psychology of prediction. *Psychological Review, 80*(4), 237–251.

Kahneman, D., & Tversky, A. (1979). Prospect theory: An analysis of decision under risk. *Econometrica, 47,* 263–291.

Kahneman, D., & Tversky, A. (1982). The simulation heuristic. In D. Kahneman, P. Slovic, & A. Tversky (Eds.), *Judgment under uncertainty: Heuristics and biases* (pp. 201–208). Cambridge, MA: Cambridge University Press.

Kasemir, B., Behringer, J., De Marchi, B., Deuker, C., Dürrenberger, G., Funtowicz, S., Gerger, A., Giaoutzi, M., Haffner, Y., Nilsson, M., Querol, C., Schüle, R., Tabara, D., Van Asselt, M., Vassilarou, D., Willi, N., & Jaeger, C. C. (1997). *Focus groups in integrated assessment: The ULYSSES pilot experience* (ULYSSES Report No. WP-97-4). Darmstadt University of Technology, Center for Interdisciplinary Studies in Technology.

Kasemir, B., & Jaeger, C. C. (1997). Introduction. In B. Kasemir, J. Behringer, B. De Marchi, C. Deuker, G. Dürrenberger, S. Funtowicz, A. Gerger, M. Giaoutzi, Y. Haffner, M. Nilsson, C. Querol, R. Schüle, D. Tabara, M. Van Asselt, D. Vassilarou, N. Willi, & C. C. Jaeger (Eds.), *Focus groups in integrated assessment: The ULYSSES pilot experience* (ULYSSES Report No. WP-97-4, p. 1). Darmstadt University of Technology.

Kasemir, B., Van Asselt, M., Deuker, C., Haffner, Y., & Willi, N. (1997). "One goes through a change while being here": A five-session IA-focus group using the TARGETS model. In B. Kasemir, J. Behringer, B. De Marchi, C. Deuker, G. Dürrenberger, S. Funtowicz, A. Gerger, M. Giaoutzi, Y. Haffner, M. Nilsson, C. Querol, R. Schüle, D. Tabara, M. Van Asselt, D. Vassilarou, N. Willi, & C. C. Jaeger (Eds.), *Focus groups in integrated assessment: The ULYSSES pilot experience* (ULYSSES Report No. WP-97-4, pp. 71–80). Darmstadt University of Technology.

Kasemir, B., Van Asselt, M., Dürrenberger, G., & Jaeger, C. C. (1999). Integrated assessment of sustainable development: Multiple perspectives in interaction. *International Journal of Environment and Pollution, 11*(4), 407–425.

Kelley, H. H. (1967). Attribution in social interaction. In D. Levine (Ed.), *Nebraska symposium on motivation* (pp. 192–238). Lincoln: University of Nebraska Press.

Kelley, H. H. (1973). The processes of causal attribution. *American Psychologist, 28,* 107–128.

Kerr, R. (1992). Unmasking a shifty climate system. *Science, 255,* 1507–1509.

Kerr, R. (1993). El Niño metamorphosis throws forecasters. *Science, 262,* 656–657.

Keynes, J. M. (1936/1964). *The general theory of employment, interest and money.* London: Hartcourt Brace.

Kleindorfer, P. R., Kunreuther, H. C., & Schoemaker, P. J. H. (1993). *Decision sciences.* Cambridge, UK: Cambridge University Press.

Kondratyev, K. Y., & Cracknell, A. P. (1998). *Observing global climate change.* London: Taylor & Francis.

Krueger, R. A. (1988). *Focus groups.* Newbury Park: Sage.

Kuhn, T. S. (1962). *The structure of scientific revolutions*. Chicago: The University of Chicago Press.
Langer, E. J. (1983). *The psychology of control*. Beverly Hills, CA: Sage.
Larson, M. S. (1977). *The rise of professionalism*. Berkeley, CA: University of California Press.
Leeson, N., & Whitley, E. (1996). *Rogue trader*. New York: Little, Brown.
Lesgold, A., Rubinson, H., Feltovich, P., Glaser, R., Klopfer, D., & Wang, Y. (1988). Expertise in a complex skill: Diagnosing X-ray pictures. In M. T. H. Chi, R. Glaser, & M. J. Farr (Eds.), *The nature of expertise* (pp. 311–342). Hillsdale, NJ: Lawrence Erlbaum Associates.
Lesgold, A. M. (1984). Acquiring expertise. In J. R. Anderson & S. M. Kosslyn (Eds.), *Tutorials in learning and memory* (pp. 31–60). San Francisco: Freeman.
Levine, J. M., Resnick, L. B., & Higgins, E. T. (1993). Social foundations of cognition. *Annual Review of Psychology*, *44*, 585–612.
Levy, H., & Markowitz, H. M. (1979). Approximating expected utility by a function of mean and variance. *The American Economic Review*, *69*(3), 308–317.
Levy, H., & Sarnat, M. (1984). *Portfolio and investment selection*. Englewood Cliffs, NJ: Prentice-Hall.
Lewis, J. D., & Weigert, A. (1985). Trust as social reality. *Social Forces*, *63*(4), 967–985.
Loomes, G. (1997, August). *Private values and public policy*. Paper presented at SPUDM 16 (Conference on Subjective Probability Utility and Decison Making), Leeds.
Luce, R., & Raiffa, H. (1957). *Games and decisions*. New York: Wiley.
Luhmann, N. (1971). *Politische Planung* [Political planning]. Opladen: Westdeutscher Verlag.
Luhmann, N. (1973). *Vertrauen* (2. Aufl.) [Trust]. Stuttgart: Enke.
Luhmann, N. (1988). Familiarity, confidence, trust. In D. G. Gambetta (Eds.), *Trust* (pp. 94–107). New York: Basil Blackwell.
Luhmann, N. (1993). *Risk: A sociological theory* (R. Barrett, Trans.). New York: Aldine de Gruyter. (German original in 1991)
MacKenzie, D. (1998). The certainty trough. In R. Williams, W. Faulkner, & J. Fleck (Eds.), *Exploring expertise* (pp. 325–329). Houndmills, UK: Macmillan.
Makridakis, S. G. (1990). *Forecasting, planning, and strategy for the 21st century*. New York: The Free Press.
March, J. G., & Simon, H. A. (1958/1993). *Organizations* (2nd ed.). Cambridge, MA: Blackwell.
Margolis, H. (1996). *Dealing with risk*. Chicago: The University of Chicago Press.
Markowitz, H. (1952). Portfolio selection. *Journal of Finance*, *7*, 77–91.
Markowitz, H. (1959). *Portfolio selection: Efficient diversification of investment*. New York: Wiley.
Markowitz, H. (1991). Foundations of portfolio theory. *Journal of Finance*, *46*, 469–477.
Martens, P., & Rotmans, J. (Eds.). (1999). *Climate change: An integrative perspective*. Dordrecht: Kluwer.
Martin, B. (Ed.). (1996). *Confronting the experts*. Albany: State University of New York.
Maturi, R. J. (1995). *Main Street beats Wall Street*. Chicago: Probus.
Mayer, R. C., Davis, J. H., & Schoorman, F. D. (1995). An integrative model of organizational trust. *Academy of Management Review*, *20*(3), 709–734.
McBean, G., & McCarthy, J. (1990). Narrowing the uncertainties. In J. T. Houghton, G. J. Jenkins, & J. J. Ephraums (Eds.), *Climate change: The IPCC scientific assessment* (pp. 311–328). Cambridge, UK: University Press.
McBean, G., Liss, P. S., & Schneider, S. H. (1996). Advancing our understanding. In J. T. Houghton, L. G. Meira Filho, B. A. Callander, N. Harris, A. Kattenberg, & K. Maskell (Eds.), *Climate change 1995: The science of climate change* (pp. 517–531). Cambridge, UK: University Press.
McCarthy, J. (1956). The inversion of functions defined by Turing machines. In D. E. Shannon & J. McCarthy (Eds.), *Automata studies. Annals of mathematical studies* (pp. 177–181). Princeton, NJ: Princeton University Press.
McClelland, G. H., Schulze, W. D., & Coursey, D. L. (1993). Insurance for low-probability hazards: A bimodal response to unlikely events. *Journal of Risk and Uncertainty*, *7*, 95–116.

McCloskey, D. N. (1990). *If you're so smart: The narrative of economic expertise.* Chicago: The University of Chicago Press.

Mead, G. H. (1924/1925). The genesis of the self and social control. *International Journal of Ethics, 35,* 251–277.

Mead, G. H. (1934). *Mind, self and society* (ed. by Charles W. Morris). Chicago: The University of Chicago Press.

Mead, G. H. (1977). *Selected papers* (ed. by Anselm Strauss). Chicago: The University of Chicago Press.

Meadows, D. H., Meadows, D. L., Randers, J., & Behrens, W. W. (1972). *The limits to growth. A report to the Club of Rome.* New York: Universe Books and Earth Island.

Medcof, J. W. (1990). PEAT: An integral model of attribution processes. In M. P. Zanna (Ed.), *Advances in experimental social psychology* (pp. 111–209). New York: Academic Press.

Meehl, P. E. (1954). *Clinical versus statistical prediction.* Minneapolis: University of Minnesota Press.

Mendelsohn, R., & Neumann, J. E. (Eds.). (1999). *The impact of climate change on the United States economy.* Cambridge, UK: Cambridge University Press.

Metcalfe, J. (1993). Novelty monitoring, metacognition, and control in a composite holographic associative recall model: Implications for Korsakoff amnesia. *Psychological Review, 100*(1), 3–22.

Michotte, A. (1946/1963). *The perception of causality.* New York: Basic Books. (Original work in French, published 1946)

Mieg, H. A. (1991). Öffentliche Moral - Illusion, Fiktion und Mythos [Public morality–illusion, fiction, myths]. In G. Roellecke (Hrsg.), *Öffentliche Moral* (S. 159–173). Heidelberg: C. F. Müller.

Mieg, H. A. (1993). *Computers as experts?* Frankfurt/New York: Lang.

Mieg, H. A. (1994). *Verantwortung: Moralische Motivation und die Bewältigung sozialer Komplexität* [Responsibility: Moral motivation and coping with social complexity]. Opladen: Westdeutscher Verlag.

Mieg, H. A. (1996). Managing the interfaces between science, industry, and education: Case studies for environment, education, and knowledge integration at the Swiss Federal Institute of Technology. *World Congress of Engineering Educators and Industry Leaders* (Vol I, pp. 529–533). Paris: UNESCO.

Mieg, H. A. (1998). The Swiss market for professional environmental services: A test of Abbott's (1988) abstraction criteria for professionalization. In A. Brosziewski & C. Maeder (Hrsg.), *Organisation und Profession* (pp. 100–122). Rorschach/St. Gallen: HFS Ostschweiz, Universität St. Gallen (Universitätsdruck).

Mieg, H. A. (2000). University-based projects for local sustainable development: Designing expert roles and collective reasoning. *International Journal of Sustainability in Higher Education, 1,* 67–82.

Mieg, H. A., Scholz, R. W., & Stünzi, J. (1996). Das Prinzip der modularen Integration: Neue Wege von Führung und Wissensintegration im Management von Umweltprojekten [The principle of modular integration: New ways of knowledge integration in environmental project management]. *Organisationsentwicklung, 15*(2), 4–15.

Milgram, S. (1963). Behavioral study of obedience. *Journal of Abnormal Psychology, 67,* 371–378.

Milgram, S. (1974). *Obedience to authority: An experimental view.* New York: Harper & Row.

Miller, G. A. (1956). The magical number seven, plus or minus two. *Psychological Review, 63,* 81–97.

Mintzberg, H. (1994). *The rise and fall of strategic planning.* New York: Prentice-Hall.

Mischel, W. (1968). *Personality and assessment.* New York: Wiley.

Morgan, M. G., & Keith, D. W. (1995). Subjective judgments by climate experts. *Environmental Policy Analysis, 29*(10), 468–476.

Mueller, C. B., & Kirkpatrick, L. C. (1993). *Federal rules of evidence: With advisory committee notes and legislative history.* Boston: Little, Brown.

Mullen, B., & Goethals, G. R. (Eds.). (1987). *Theories of group behavior*. New York: Springer.

Musen, M. A., & van der Lei, J. (1989). Knowledge engineering for clinical consulting programs: Modelling the application area. *Methods of Information in Medicine, 28*(1), 28–35.

Neukom, J. G. (1975). *McKinsey memoirs*. Chicago: John G. Neukom.

Newell, A. (1982). The knowledge level. *Artificial Intelligence, 18*, 87–127.

Newell, A., & Simon, H. A. (1972). *Human problem solving*. Englewood Cliffs, NJ: Prentice-Hall.

Nisbett, R. E., & DeCamp Wilson, T. (1977). Telling more than we can know: Verbal reports on mental processes. *Psychological Review, 84*(3), 231–259.

Nordhaus, W. D. (1994, February). Expert opinion on climate change. *American Scientist, 82*, 45–51.

Nowotny, H. (1979). *Kernernergie: Gefahr oder Notwendigkeit* [Nuclear power: Danger or necessity]. Frankfurt: Suhrkamp.

Ochs, E. (1991). Socialization through language and interaction: A theoretical introduction. *Issues in Applied Linguistics, 2*(2), 143–147.

OECD. (Ed.). (1991). *Systemic risks in security markets*. Paris: Author.

Oskamp, S. (1965). Overconfidence in case-study judgments. *Journal of Consulting Psychology, 29*, 261–265.

Oskamp, S. (1967). Clinical judgment from the MMPI: Simple or complex? *Journal of Clinical Psychology, 23*, 411–415.

Otway, H., & von Winterfeldt, D. (1992). Expert judgment in risk analysis and management: Process, context, and pitfalls. *Risk Analysis, 12*(1), 83–93.

Pahl-Wostl, C. (1995). *The dynamic nature of ecosystems*. Chichester: Wiley.

Park, R. E., Burgess, E. W., & McKenzie, R. D. (1925/1967). *The city*. Chicago: University of Chicago Press.

Parsons, T. (1939). The professions and social structure. *Social Forces, 17*, 457–467.

Parsons, T. (1951). *The social system*. Glencoe, IL: The Free Press.

Parsons, T. (1968). Professions. *International Encyclopedia of the Social Sciences, 12*, 536–547.

Patel, V. L., & Ramoni, M. E. (1997). Cognitive models of directional inference in expert medical reasoning. In P. J. Feltovich, K. M. Ford, & R. R. Hoffman (Eds.), *Expertise in context: Human and machine* (pp. 67–99). Menlo Park, CA: AAAI.

Peters, H. P. (1995). The interaction of journalists and scientific experts: Co-operation and conflict between two professsional cultures. *Media, Culture & Society, 17*, 31–48.

Peters, Th. J., & Waterman, R. H. (1982). *In search of excellence*. New York: Harper & Row.

Popper, K. (1934/1959). *The logic of scientific discovery*. New York: Harper & Row.

Posner, M. I. (1988). Introduction: What is it to be an expert? In M. T. H. Chi, R. Glaser, & M. J. Farr (Eds.), *The nature of expertise* (pp. xxix–xxxvi). Hillsdale, NJ: Lawrence Erlbaum Associates.

Price, D. K. (1965). Scientific illiteracy and democratic theory. *Daedalus, 112*(2), 49–64.

Pritchett, W. K. (1974). The Greek state at war (Part II). Berkeley, CA: The University of California Press.

Prognos, A. G. (1996). *Energieperspektiven der Szenarien I bis III 1990–2030: Synthesebericht* [Energy perspectives for the scenarios I to III 1990–2030: a synthesis report]. Basel: Bundesamt für Energiewirtschaft.

Putnam, H. (1983). Why there isn't a ready-made world. In H. Putnam (Ed.), *Realism and reason* (pp. 205–228). Cambridge, MA: Cambridge University Press.

Rescher, N. (1989). *Cognitive economy*. Pittsburgh, PA: The University of Pittsburgh Press.

Rescher, N. (1988). *Rationality*. Oxford: Clarendon.

Ringland, G. (1998). *Scenario planning: Managing for the future*. Chichester: Wiley.

Ross, L. (1977). The intuitive psychologist and his shortcomings: Distortions in the attribution process. In L. Berkowitz (Ed.), *Advances in experimental social psychology* (pp. 173–220). New York: Academic Press.

Rossi, F. F. (Ed.). (1991). *Expert witnesses*. Chicago: American Bar Association.

Rotmans, J., & de Vries, H. J. M. (1997). *Perspectives on global change: The TARGETS approach.* Cambridge, UK: Cambridge University Press.

Rotmans, J., & van Asselt, M. (1999). Integrated assessment modelling. In P. Martens & J. Rotmans (Eds.), *Climate change: An integrative perspective* (pp. 239–275). Dordrecht: Kluwer.

Rueschemeyer, D. (1983). Professional autonomy and the social control of expertise. In R. Dingwall & P. Lewis (Eds.), *The sociology of the professions* (pp. 38–58). London: Macmillan.

Rueschemeyer, D. (1986). *Power and the division of labour.* Cambridge, UK: Polity.

Sarin, R. K., & Weber, M. (1993). Effects of ambiguity in market experiments. *Management Science, 39*(5), 602–615.

Schegloff, E. A. (1991). Reflections on talk and social structure. In D. Boden & D. Zimmerman (Eds.), *Talk and social structure* (pp. 44–70). Cambridge, UK: Polity.

Scholz, R. W. (1983). Biases, fallacies, and the development of decision making. In R. W. Scholz (Ed.), *Decision making under uncertainty* (pp. 3–18). Amsterdam: North-Holland.

Scholz, R. W., Bösch, S., Mieg, H. A., & Stünzi, J. (Hrsg.). (1997). *Zentrum Zürich Nord, Fallstudie 1996* [City center "Zurich North" Case Study 1996]. Zürich: Verlag der Fachvereine.

Scholz, R. W., Mieg, H. A., & Weber, O. (1997). Mastering the complexity of environmental problem solving by case study approach. In R. W. Scholz & A. C. Zimmer (Eds.), *Qualitative aspects of decision making* (pp. 169–186). Lengerich: Pabst.

Scholz, R. W., & Tietje, O. W. (in press). *Embedded case study methods.* Thousand Oaks: Sage.

Schön, D. A. (1991). *The reflective practitioner.* Aldershot Hants, GB: Avebury.

Schwager, J. D. (1989). *Market wizards: Interviews with top traders.* New York: New York Institute of Finance.

Schwager, J. D. (1992). *The new market wizards: Conversations with America's top traders.* New York: Harper Business.

Scribner, S. (1984). Studying working intelligence. In B. Rogoff & J. Lave (Eds.), *Everyday cognition: Its development in social context* (pp. 9–40). Cambridge, MA: Harvard University Press.

Semin, G. R., & Manstead, A. S. R. (1983). *The accountability of conduct.* London: Academic Press.

Shalin, V. L., Geddes, N. D., Bertram, D., Szczepkowski, M. A., & DuBois, D. (1997). Expertise in dynamic, physical task domains. In P. J. Feltovich, K. M. Ford, & R. R. Hoffman (Eds.), *Expertise in context: Human and machine* (pp. 195–217). Menlo Park, CA: AAAI.

Shannon, C. E., & Weaver, W. (1949). *The mathematical theory of communication.* Urbana: University of Illinois Press.

Shanteau, J. (1992a). Competence in experts: The role of task characteristics. *Organizational Behavior and Human Decision Processes, 53,* 252–266.

Shanteau, J. (1992b). The psychology of experts. In G. Wright & F. Bolger (Eds.), *Expertise and decision support* (pp. 11–23). New York: Plenum.

Shanteau, J., & Stewart, T. R. (1992). Why study expert decision making? Some historical perspectives and comments. *Organizational Behavior and Human Decision Processes, 53,* 95–106.

Sharpe, W. F. (1964). Capital asset prices: A theory of market equilibrium under conditions of risk. *Journal of Finance, 19,* 425–442.

Sharpe, W. F., Alexander, G. J., & Bailey, J. V. (1995). *Investments* (5th ed.). Englewood Cliffs, NJ: Prentice-Hall.

Shefrin, H. (1999). *Beyond greed and fear: Understanding behavioral finance and the psychology of investing.* Boston, MA: Harvard Business School Press.

Shepherd, J. C. (1973). Relations with the expert witnesses. In G. W. Holmes (Ed.), *Experts in litigation* (pp. 19–24). Ann Arbor: Institute of Continuing Legal Education.

Shiller, R. J. (1981). Do stock prices move too much to be justified by subsequent changes in dividend? *The American Economic Review, 71*(3), 421–436.

Shiller, R. J. (1990). Speculative prices and popular models. *Journal of Economic Perspectives, 4*(2), 55–65.

Shiller, R. J. (1991). *Market volatility* (3rd ed.). Cambridge, MA: MIT Press.

Shiller, R. J. (2000). *Irrational exuberance.* Princeton, NJ: Princeton University Press.

Shortliffe, E. H. (1989). Testing reality: The introduction of decision-support technologies for physicians. *Methods of Information in Medicine, 28*(1), 1–5.

Shortliffe, E. H., & Buchanan, B. G. (1984). A model of inexact reasoning in medicine. In B. G. Buchanan & E. H. Shortliffe (Eds.), *Rule-based expert systems* (pp. 233–262). Reading, MA: Addison-Wesley.

Shulman, L. S., & Carey, N. B. (1984). Psychology and the limitations of individual rationality: Implications for the study of reasoning and civility. *Review of Educational Research, 54*(4), 501–524.

Simmel, G. (1907). *Philosophie des Geldes* (2. Aufl.) [The philosophy of money]. Leipzig: Duncker & Humblot.

Simmel, G. (1908). *Soziologie: Untersuchungen über die Formen der Vergesellschaftung* [Sociology: Studies on the forms of sociation]. Leipzig: Duncker & Humblot.

Simmel, G. (1971). *On individuality and social forms* (K. H. Wolff, Trans., D. N. Levine, Ed.). Chicago: The University of Chicago Press.

Simon, H. A. (1956). Rational choice and the structure of environment. *Psychological Review, 63*(2), 129–138.

Simon, H. A. (1972). Theories of bounded rationality. In C. B. Radner & R. Radner (Eds.), *Decision and organization* (pp. 161–176). Amsterdam: North-Holland.

Simon, H. A. (1976). From substantive to procedural rationality. In S. J. Latsis (Ed.), *Methods and appraisal in economics* (pp. 129–148). Cambridge, MA: Cambridge University Press.

Simon, H. A. (1979). *Models of thought.* New Haven: Yale University Press.

Simon, H. A. (1982). *Models of bounded rationality: Behavioral economics and business organization.* Cambridge, MA: The MIT Press.

Simon, H. A. (1990). Invariants of human behavior. *Annual Review of Psychology, 41*, 1–19.

Simon, H. A. (1992). Thinking by computers (1966). In H. A. Simon, M. Egidi, R. Marris, & R. Viale (Eds.), *Economics, bounded rationality and the cognitive revolution* (pp. 55–75). Brookfield: Elgar.

Simonton, D. K. (1988). *Scientific genius.* Cambridge, UK: Cambridge University Press.

Slovic, P. (1987). Perception of risk. *Science, 236*, 280–285.

Sorensen, J. T., & Sorensen, T. L. (1974). The conflict of professionals in bureaucratic organizations. *Administrative Science Quarterly, 19*, 98–106.

Soros, G. (1994). *The alchemy of finance.* New York: Wiley.

Soros, G. (1998). *The crisis of global capitalism.* New York: Public Affairs.

Stael von Holstein, C.-A. S. (1972). Probabilistic forecasting: An experiment related to the stock market. *Organizational Behavior and Human Performance, 8*, 139–158.

Stasser, G. (1992a). Information salience and the discovery of hidden profiles by decision-making groups: A "thought experiment." *Organizational Behavior and Human Decision Processes, 52*, 156–181.

Stasser, G. (1992b). Pooling of unshared information during group discussion. In S. Worchel, W. Wood, & J. A. Simpson (Eds.), *Group process and productivity.* Newbury Park, CA: Sage.

Stasser, G., & Stewart, D. D. (1992). Discovery of hidden profiles by decision-making groups: Solving a problem versus making a judgment. *Journal of Personality and Social Psychology, 68*(3), 426–434.

Stasser, G., Stewart, D. D., & Wittenbaum, G. D. (1995). Expert roles and information exchange during discussion: The importance of knowing who knows what. *Journal of Experimental Social Psychology, 31*, 244–265.

Stasser, G., & Titus, W. (1985). Effects of information load and percentage of shared information on the dissemination of unshared information during group discussion. *Journal of Personality and Social Psychology, 53*(1), 81–93.

Stegmüller, W. (1979). *The structuralist view of theories.* Berlin: Springer.

Stein, E. W. (1992). A method to identify candidates for knowledge acquisition. *Journal of Management Information Systems, 9*, 161–178.

Stein, E. W. (1997). A look at expertise from a social perspective. In P. J. Feltovich, K. M. Ford, & R. R. Hoffman (Eds.), *Expertise in context: Human and machine* (pp. 181–194). Menlo Park, CA: AAAI.

Stewart, D. D., & Stasser, G. (1995). Expert role assignment and information sampling during collective recall and decision making. *Journal of Personality and Social Psychology, 69*(4), 619–628.

Stickel, S. E. (1990). Predicting individual analyst earnings forecasts. *Journal of Accounting Research, 28*(2), 409–417.

Stickel, S. E. (1991). Common stock returns surrounding earnings forecast revisions: More puzzling evidence. *The Accounting Review, 66*(2), 402–416.

Stigler, G. J. (1942/1966). *The theory of price* (4th ed.). New York: Macmillan.

Stinchcombe, A. L. (1968). *Constructing social theory.* Chicago: The University of Chicago Press.

Storms, M. D. (1973). Videotape and the attribution process: Reversing actors' and observers' points of view. *Journal of Personality and Social Psychology, 27*(2), 165–175.

Summers, H. G., Jr. (1981). *On strategy: The Vietnam war in context.* U.S. Army College: Strategic Studies Institute. (Reference from Mintzberg, 1994)

Sundali, J. A., & Atkins, A. B. (1994). Expertise in investment analysis: Fact or fiction. *Organizational Behavior and Human Decision Processes, 59,* 223–241.

Tajfel, H., & Wilkes, A. L. (1963). Classification and quantitative judgement. *British Journal of Psychology, 54*(2), 101–114.

Taylor, P. W. (1986). *Respect for nature.* Princeton: Princeton University Press.

Templeton, J. F. (1994). *The focus group.* Chicago: Probus.

Tetlock, P. E. (1992). The impact of accountability on judgment and choice: Toward a social contingency model. In M. P. Zanna (Ed.), *Advances in experimental social psychology* (pp. 331–376). New York: Academic Press.

Tetlock, P. E., & Levi, A. (1982). Attribution biases: On the inconclusiveness of the cognition-motivation debate. *Journal of Experimental Social Psychology, 18*(68–88).

Thaler, R. H. (Ed.). (1993). *Advances in behavioral finance.* New York: Russell Sage Foundation.

The Prince of Wales' Urban Design Task Force. (1997). *Berlin 1997: Perspektivenwerkstatt Schloßplatz-Areal* [Berlin future workshop for the Schloßplatz area]. London: John Tompson & Partners.

Thompson, M., Ellis, R., & Wildavsky, A. (1990). *Cultural theory.* Boulder: Westview.

Thompson Klein, J., Grossenbacher-Mansuy, W., Häberli, R., Bill, A., Scholz, R. W., & Welti, M. (Eds.). (2001). *Transdisciplinarity: Joint problem solving among science, technology, and society.* Basel: Birkhäuser.

Trenberth, K. E., Houghton, J. T., & Meira Filho, L. G. (1996). The climate system: An overview. In J. T. Houghton, L. G. Meira Filho, B. A. Callander, N. Harris, A. Kattenberg, & K. Maskell (Eds.), *Climate change 1995: The science of climate change* (pp. 51–64). Cambridge, UK: University Press.

Trumbo, D., Adams, C., Milner, M., & Schipper, L. (1962). Reliability and accuracy in the inspection of hard red winter wheat. *Cereal Science Today, 7,* 62–71.

Tversky, A., & Kahneman, D. (1982). Judgment under uncertainty: Heuristics and biases. In D. Kahneman, P. Slovic, & A. Tversky (Eds.), *Judgment under uncertainty: Heuristics and biases* (pp. 3–20). Cambridge, MA: Cambridge University Press.

UIC (Union internationale des Chemins de fer) (1997, October). *Rail plan – scenario – strategy – action.* Paris: UIC.

UNEP (Ed.). (1996). *Handbook for the international treaties for the protection of the ozone layer.* Geneva: UNEP (United Nations Environment Programme).

UNEP/WMO (Eds.). (n.d.). *United Nations Framework Convention on Climate Change* (text). Geneva: UNEP/WMO.

van Asselt, M., & Rotmans, J. (1995). *Uncertainty in integrated assessment modelling* (GLOBO Report Series No. 9). Bilthoven (NL): National Institute of Public Health and the Environment, Research for Man and Environment.

van Asselt, M., & Rotmans, J. (1996). Uncertainty in perspective. *Gobal Environmental Change, 6*(2), 121–157.

Vlek, C., & Stallen, P.-J. (1980). Rational and personal aspects of risks. *Acta Psychologica, 45*, 273–300.

von Neumann, J., & Morgenstern, O. (1944/1972). *Theory of games and economic behavior*. Princeton, NJ: Princeton University Press.

von Winterfeldt, D., & Edwards, W. (1986). *Decision analysis and behavioral research*. New York: Cambridge University Press.

von Wright, G. H. (1971). *Explanation and understanding*. Ithaca, NY: Cornell University Press.

Wason, P. C., & Johnson-Laird, P. N. (1972). *Psychology of reasoning*. Cambridge, MA: Harvard University Press.

Wates, N. (1996). *Action planning*. Hastings: Prince of Wales Institute of Architecture.

Watson, A. S. (1973). On getting the jury to trust the expert. In G. W. Holmes (Ed.), *Experts in litigation* (pp. 75–82). Ann Arbor: Institute of Continuing Legal Education.

Weber, M. (1946). *Essays in sociology* (H. Gerth & C. Wright Mills, Trans.). New York: Oxford University Press.

Weber, M. (1947/64). *Social and economic organization* (T. Parsons, A. M. Henderson, & T. Parsons, Trans.). New York: The Free Press.

Weber, M. (1972). *Wirtschaft und Gesellschaft* (5., rev. Aufl., hrsg. v. J. Winckelmann) [Economy and society]. Tübingen: Mohr.

Weber, M. (1979). *Economy and Society* (Vol. I, G. Roth & C. Wittich, Trans.). Berkeley, CA: University of California Press.

Weber, M., & Camerer, C. (1998). The disposition effect in securities trading: An experimental analysis. *Journal of Economic Behavior and Organization, 33*, 167–184.

Wegner, D. M. (1987). Transactive memory: A contemporary analysis of the group mind. In B. Mullen & G. R. Goethals (Eds.), *Theories of group behavior* (pp. 185–208). New York: Springer.

Weiner, B. (1986). *An attributional theory of motivation and control*. New York: Springer.

Weiner, B. (1995). Inferences of responsibility and social motivation. In M. P. Zanna (Ed.), *Advances in experimental social psychology* (pp. 1–47). New York: Academic Press.

Weizenbaum, J. (1976). *Computer power and human reason*. San Francisco: Freeman.

Wellmann, C. R. (1988). Die Pflichten des Sachverständigen und seine zivilrechtliche Verantwortung [Duties of experts and their civil liability]. In C. R. Wellmann, *Der Sachverständige in der Praxis* (pp. 1–33). Düsseldorf: Werner.

Wildavsky, A. (1974). *The politics of the budgetary process* (2nd ed.). Boston, MA: Little, Brown.

Williams, R., Faulkner, W., & Fleck, J. (Eds.). (1998). *Exploring expertise*. Houndmills, UK: Macmillan Press.

Winograd, T., & Flores, F. (1986). *Understanding computers and cognition*. Norwood, NJ: Ablex.

Wittgenstein, L. (n.d.). *Philosophical investigations* (G. E. M. Anscombe, Trans.). (As to the bibliographical history of this book, see, e.g., Baker, G. P., & Hacker, P. M. S. [1980], *Understanding and meaning*. Oxford: Basil Blackwell.)

Wood, C. (1988). *Boom and blust*. London: Sidgwick & Jackson.

Wright G., & Bolger, F. (Eds.). (1992). *Expertise and decision support*. New York: Plenum.

Yates, F. J. (1990). *Judgment and decision making*. Englewood Cliffs, NJ: Prentice-Hall.

Yates, F. J. (Ed.). (1992). *Risk-taking behavior*. Chichester: Wiley.

Yates, F. J., & Stone, E. (1992). The risk construct. In F. J. Yates (Ed.), *Risk-taking behavior* (pp. 1–25). Chichester: Wiley.

Yates, J. F., McDaniel, L. S., & Brown, E. S. (1991). Probabilistic forecasts of stock prices and earnings: The hazards of nascent expertise. *Organizational Behavior and Human Decision Processes, 49*, 60–79.

Yu, V. L., Fagan, L. M., Wraith Bennett, S., Clancey, W. J., Carlisle Scott, A., Hannigan, J. F., Buchanan, B. G., & Cohen, S. N. (1984). A model of inexact reasoning in medicine. In B. G. Buchanan & E. H. Shortliffe (Eds.), *Rule-based expert systems* (pp. 589–596). Reading: Addison-Wesley.

Zey, M. (1993). *Banking on fraud*. Hawthorne: de Gruyter.

Author Index

A

B

C

H

I

J

K

L

Subject Index

D

E

H

I

J

K

L